£ 25.00

NEUTRAL MODELS IN BIOLOGY

NEUTRAL MODELS IN BIOLOGY

Edited by
MATTHEW H. NITECKI
Field Museum of Natural History

ANTONI HOFFMAN
Polish Academy of Sciences

New York   Oxford
OXFORD UNIVERSITY PRESS
1987

Oxford University Press

Oxford New York Toronto
Delhi Bombay Calcutta Madras Karachi
Petaling Jaya Singapore Hong Kong Tokyo
Nairobi Dar es Salaam Cape Town
Melbourne Auckland

and associated companies in
Beirut Berlin Ibadan Nicosia

Copyright © 1987 by Oxford University Press, Inc.

Published by Oxford University Press, Inc.,
200 Madison Avenue, New York, New York 10016

Oxford is a registered trademark of Oxford University Press

All rights reserved. No part of this publication may be reproduced,
stored in a retrieval system, or transmitted, in any form or by any means,
electronic, mechanical, photocopying, recording, or otherwise,
without the prior permission of Oxford University Press.

Library of Congress Cataloging-in-Publication Data
Neutral models in biology.
Bibliography: p. Includes index.
1. Biological models. I. Nitecki, Matthew H. II. Hoffman, Antoni.
QH324.8.N47 1987   574'.07'24   86-33235
ISBN 0-19-505099-1

9 8 7 6 5 4 3 2 1

Printed in the United States of America
on acid-free paper

This book is dedicated
to the memory of
THOMAS J. M. SCHOPF
1939–1984

Preface

Biological theories explain the observed patterns of life at various levels of organization, from molecular to ecological, and on various time scales, from physiological to geological. The goal is approached by proposing general patterns under reasonable boundary conditions. There is, however, more than just one theoretical model that can account for empirical data and, therefore, model assumptions must also be tested. It is at this point that neutral models inescapably enter the argument. For if a model could be proposed which accounts for the pattern and yet involves no elaborate theory at all, and which follows from the principle of randomness under only a minimum number of assumptions that are unquestionable, why search for more sophisticated theories? The criterion of parsimony suggests that the neutral model should then be accepted as the explanation. But there is virtually no pattern in nature that could not be obtained by some stochastic process, more or less complex, elegant or even plausible. What is, then, the appropriate methodology of testing neutral models? What is their use in the life sciences?

This debate has raged for well over a decade, although it has only recently become "hot" and spectacular. Our book brings together philosophers and leading scientists from various biological disciplines to promote a mutual understanding, if not a consensus, on this fundamental but highly diverse methodological issue.

The present volume is the result of the Eighth Spring Systematics Symposium held in May of 1985 at the Field Museum of Natural History in Chicago, and supported by the National Science Foundation grant (BSR-8415663). We are grateful to the National Science Foundation for its generous financial support.

As in most of the previous publications resulting from the Field Museum Symposia, we have been fortunate in having the assistance of Doris V. Nitecki, who helped patiently and efficiently with all editorial matters, particularly with editing, proofreading, and preparation of indexes.

May 1987   M.H.N.
                                                                                                   A.H.

Contents

Contributors	x
Introduction: Neutral Models as a Biological Research Strategy ANTONI HOFFMAN AND MATTHEW H. NITECKI	3

I. MOLECULAR AND GENETIC MODELS

1. Neutral Models in Molecular Evolution — 11
   JAMES F. CROW
2. False Models as Means to Truer Theories — 23
   WILLIAM C. WIMSATT
3. Self-Organization, Selective Adaptation, and Its Limits:
   A New Pattern of Inference in Evolution and Development — 56
   STUART A. KAUFFMAN

II. ECOLOGICAL MODELS

4. How to Be Objective in Community Studies — 93
   L. B. SLOBODKIN
5. On the Use of Null Hypotheses in Biogeography — 109
   PAUL H. HARVEY

III. PALEONTOLOGICAL MODELS

6. Neutral Models in Paleobiology — 121
   DAVID M. RAUP
7. Neutral Model of Taxonomic Diversification in the Phanerozoic:
   A Methodological Discussion — 133
   ANTONI HOFFMAN
8. Testing Hypotheses or Fitting Models?
   Another Look at Mass Extinctions — 147
   STEPHEN M. STIGLER

Index — 160

Contributors

James F. Crow. Genetics Department, University of Wisconsin, Madison, Wisconsin 53706

Paul H. Harvey. Department of Zoology, University of Oxford, South Parks Road, Oxford OX1 3PS, England

Antoni Hoffman. Institute of Paleobiology, Polish Academy of Sciences, Al. Żwirki i Wigury 93, PL-02-089 Warszawa, Poland

Stuart A. Kauffman. Department of Biochemistry and Biophysics, University of Pennsylvania, Philadelphia, Pennsylvania 19104

Matthew H. Nitecki. Department of Geology, Field Museum of Natural History, Roosevelt Road at Lake Shore Drive, Chicago, Illinois 60605

David M. Raup. Department of Geophysical Sciences, University of Chicago, Chicago, Illinois 60637

L. B. Slobodkin. Ecology and Evolution, State University of New York at Stony Brook, Stony Brook, New York 11794

Stephen M. Stigler. Department of Statistics, University of Chicago, Chicago, Illinois 60637

William C. Wimsatt. Department of Philosophy, University of Chicago, Chicago, Illinois 60637

NEUTRAL MODELS IN BIOLOGY

Introduction: Neutral Models as a Biological Research Strategy

ANTONI HOFFMAN and MATTHEW H. NITECKI

The term "neutral model" refers to a research strategy employed throughout a wide range of biological sciences—from molecular biology to biogeography and paleontology. Its precise meaning, however—and hence the aims and means of this research strategy—is obfuscated by the variation of its usage in the different biological disciplines. "Neutral model" belongs to a whole family of terms, which are sometimes, but by far not universally, used interchangeably; these are "null hypothesis," "null model," "random model," "baseline model," "stochastic approach," "neutral theory," etc. A number of these terms are in fact employed, often in different contexts and with different meanings, in the chapters of this book.

We suspect that this terminological ambiguity considerably contributes to the often hot and indeed passionate debates ranging in so many biological disciplines. A clarification of this semantic tangle, while certainly most advantageous to biology, would, however, be a rather pedantic task. Therefore, instead of attempting to disentangle the complex network of concepts designated by this family of terms, we will focus on the biological research strategy which calls for the application of neutral models and the associated methodological tools.

In this volume, several particular neutral models are discussed in the context of a variety of biological disciplines—various branches of genetics (Chapters 1–3), ecology and biogeography (Chapters 4 and 5), and paleontology (Chapters 6–8). This broad scope of the book reflects the generality of the research strategy that prompts biologists to develop neutral models. Models in science always serve the purpose of approaching the explanation of some empirical phenomena. What is, then, specific about neutral models? Why are they so widely employed in biological sciences? What kind of problems are being addressed by biologists while formulating and evaluating these models? In order to address these questions, we have first to discuss briefly the biological mode of explanation.

Biological inquiries describe the various phenomena of life and their history, and aim to explain them. Explanation in science is always causal. It means that the phenomenon to be explained (*explanandum*) is logically deducible from a set of general laws, or theory, accompanied by particular statements specifying the initial and boundary conditions (*explanans*). Explanation thus provides a set of sufficient conditions for the occurrence of the phenomenon or class of phenomena under consideration. This is Hempel and Oppenheim's (1948) "nomologico-deductive" pattern of scientific explanation. In the historical context, it indicates a causal process that may have resulted in the observed event or sequence of events.

The first step in any biological endeavor therefore must be the identification and the adequate description of the phenomenon to be explained. Depending on the nature of the problem addressed, the biologist must describe the structure of the system, or the web of interactions within the system as well as its links with the environment, or the pattern of change of the system in space and/or time. Thus, a functional morphologist may have to first describe the anatomy of an individual sponge; an ecologist, the trophic web and energy flow in a lake community; a paleontologist, the pattern or morphologic evolution in an extinct ostracod species. Such description is by no means a simple task. Every biologist can enumerate a long list of particular technical obstacles to be overcome in various biological disciplines.

Difficulties arise, however, not only from particular technicalities but also from a more fundamental problem; when description is a step toward explanation rather than a goal in itself, it must deal with functionally meaningful entities. Our functional morphologist must somehow identify the individual sponge with all its offshoots that may fuse again; our ecologist must delimit the community with all its significant components; and the paleontologist must recognize the evolutionary species among its congeneric and perhaps very similar forms. This may be a formidable challenge.

The description in biology often can be accomplished only by hermeneutic analysis (Hoffman and Reif, submitted for publication). If there is a conceivable explanation for the existence, structure, action, and/or pattern of a given system, the system may be tentatively interpreted as a functionally meaningful entity. Observation, experimentation, and modeling can then be implemented to achieve the adequate description of the entity. These analytic procedures, however, may lead to the conclusion that the initial identification of the entity has been unsatisfactory, and may thus demand a redefinition of the scope of the study. Improving an understanding of the biological reality leads to improved description, which in turn improves the understanding.

There is in biology a hierarchy of the levels of organization (from molecules to biosphere) at which functionally meaningful entities can be described. In this descriptive sense, biology obviously is a hierarchical science (Mayr 1982). One of the most contentious issues in modern biology is whether this descriptive, or epistemological, hierarchy implies also a hierarchy of causal processes that explain biological phenomena (e.g., Salthe 1985; Hecht and Hoffman 1986). This may sound like a highly abstract and purely philosophical problem, but it has in fact direct implications for all biologists who seek to explain their observations. They must all deal with the question of how to decide whether a set of empirical phenomena from a higher level of the descriptive hierarchy is correctly explained by lower-level mechanisms, or whether some additional mechanisms should be invoked. A geneticist, for example, may address the question whether genetic drift adequately explains the differentiation between some populations, or whether natural selection should also be invoked. A paleontologist may ask whether the change in taxonomic composition of a clade over long periods of time is sufficiently explained by natural selection plus random speciation and extinction events, or whether species selection should be included in the explanation. How to decide if one-level explanations are satisfactory? What levels of the explanatory hierarchy are valid?

In the Hempel-Oppenheim model, explanation must refer to a set of general laws, or theory. Philosophers for a long time have debated the existence, and even the possibility, of biological theories. Most recently, Rosenberg (1985) argued that biology is spanned between fundamental biochemical laws and the theory of natural selection. In his view, biochemistry provides the initial conditions for the action of natural selection. Apart from these two universal statements about life phenomena, all biology is only case studies. This is simply because the taxonomic category "species" is not a class in the logical sense of the word; except for gene frequency changes, all biological phenomena are unique and, hence, not amenable to explanation by a general theory. In this view, then, the explanatory hierarchy of biology consists of two levels only.

That each species has its unique individuality does not preclude, however, that all species also have some universal properties. Such universal properties do in fact exist at all levels of the descriptive hierarchy of biological phenomena. It is therefore perfectly possible that, in addition to the biochemical theory and the theory of natural selection, several other biological theories can also be developed. For example, all organisms have a genotype and a phenotype, and therefore a general theory of epigenetic relationships is conceivable; all species are capable of interspecific competitition, and one may therefore approach a general theory in ecology; all species may speciate and become extinct, and therefore the theory of species selection is logically viable, etc. Such theories once successfully developed and established will have to be referred to in explanations concerning their respective domains. The explanatory hierarchy will then be fully congruent with the descriptive hierarchy of biological entities.

Until this happens, however, that is, for as long as the biologist is forced to grope into the unknown, the only firm basis for biological explanations is indeed provided by two laws: (1) biophysics and biochemistry, and (2) the neo-Darwinian theory of the interplay between natural selection, genetic drift, mutation pressure, and changing extrinsic environment. At each level of the descriptive hierarchy, whatever plausible general laws the biologist's invention and personal inclinations suggest may be invoked in the explanation. The methodological version of Occam's razor (called "pragmatic reductionism" by Hoffman 1983) suggests, however, that such general laws should not be accepted until they represent a necessary component of the explanatory hierarchy of biology. For example, if the pattern of species distribution on islands can be explained by a plausible combination of the consequences of the theory of natural selection and realistic auxiliary conditions, there is no need to refer to competition theory in this context.

It is a separate problem how to determine if some plausible, and hence potentially acceptable, levels of the explanatory hierarchy are indeed necessary additions to our conceptual framework of the biological reality. A variety of methodological criteria can be applied to this end and are exhaustively discussed in the individual chapters of this book. They range from heuristic usefulness (Chapter 1) and theoretical richness (Chapter 4) to statistical evaluation of the gain in the likelihood of various explanations versus the cost of additional assumptions (Chapter 8). All these criteria can be, and indeed are, widely employed by biologists whenever they have to choose between alternative models.

When such alternative models, however, span more than one level of the explanatory hierarchy, biologists face a much more fundamental dilemma because their decision may imply an increase in complexity of the explanatory, ontological hierarchy of biology. It is at this point that neutral models inescapably enter the argument. Neutral models are sufficiency models that represent phenomena at various levels of the descriptive hierarchy; and they represent them solely by reference to the established theories of those levels of the explanatory hierarchy whose validity is indisputable. In other words, neutral models are those models whose assumptions need no additional theoretical justification.

Those neutral models which represent phenomena of higher levels by reference to lower-level theories and auxiliary conditions are stochastic rather than deterministic. For example, the neo-Darwinian theory does not predict anything particular about the pattern of species distribution on islands. It only tells that this pattern should conform to a stochastic pattern resulting from random colonization and local extermination. If the predicted stochastic pattern does not fit the empirical data, competition theory may be needed to explain the rules of assembly. To test these predictions, and thus to decide if there is a need for additional, higher levels of the explanatory hierarchy, appropriate stochastic models must be developed and tested—as statistical null hypotheses—against the empirical data. Of course, this is a tremendous and highly controversial task (Chapter 5).

Under the conditions of enormous spatial and temporal variation in biological and environmental circumstances, lower-level theories about individual organisms and populations predict randomness of higher-level phenomena among species, clades, communities, and biotas. This is why neutral models of ecology, biogeography, and paleontology refer to randomness. (Note, however, that if the biological and environmental circumstances are constrained by whatever factors, the resulting higher-level patterns will also deviate from randomness.)

That a stochastic model represents well a set of empirical paleontological data, does not mean that these data do not call for explanation. To accept a stochastic model not only as description but also as explanation of paleontological data may only be justified by the option for a fundamental ontologic indeterminism (Hoffman 1981). Yet it seems hardly disputable that the fate of organisms, populations, species, and communities is causally determined, even though in a weak, statistical sense. Therefore, a fit between a stochastic model and empirical paleontological data does not mean that the data are actually explained by a random process. It only implies that the explanation by means of lower-level theories is adequate; it means that the need for a specifically paleontological theory remains to be demonstrated (Chapters 6 and 7). Of course, the usual caveats about the power of testing should be reiterated here. It is a matter of statistical considerations and methodological options to decide if the fit between neutral model and empirical pattern is sufficient to obviate the search for higher-level evolutionary forces.

The reverse is true about phenomena that occur below the level of processes considered by the neo-Darwinian theory. Neutral models of such phenomena must assume the laws of biophysics and biochemistry but not the laws of natural selection. These models may be either deterministic or stochastic. If a deterministic neutral model of such lower-level phenomena succeeds, it implies reducibility of the considered phenomena to biophysical and biochemical processes. On the other hand, natural selection is a higher-level phenomenon in this context. A higher-level process can be expected to constrain, and thus to introduce some order into, lower-level phenomena. Therefore, if a stochastic model satisfactorily represents patterns of, say, molecular evolution or genome organization, such a fit may suggest that the theory of natural selection is superfluous in the explanation; it may indicate that a neutral theory is more appropriate in this context (Chapters 1 and 3).

Thus, the application of neutral models in biology presents a paradox. If a stochastic neutral model of higher-level phenomena, above the level considered by the neo-Darwinian theory, withstands the test of empirical data, then the higher-level phenomena also belong, or at least may belong, to the explanatory domain of natural selection. If, on the other hand, a stochastic neutral model at a level below the one of natural selection cannot be rejected as a null hypothesis, this implies that the considered phenomena do not, or at least may not, belong to the explanatory domain of natural selection. This apparent paradox is a consequence of the hierarchical structure of life and, consequently, biology.

The biological research strategy using neutral models aims at analysis of phenomena at various levels of descriptive hierarchy. This is in order to determine if they can be explained by reference to the theories of the established levels of the explanatory hierarchy. Thus far, there is no evidence whatsoever for the existence of theories specific for the levels above those envisaged by neo-Darwinism. By contrast, there are good, even though perhaps still not compelling, arguments for the existence of theories specific for the levels below the level of natural selection—that is, molecular evolution and genome organizations.

It is conceivable that all levels of the descriptive, epistemological hierarchy of biological patterns and phenomena have their equivalents in the explanatory, ontological hierarchy of causal processes. This is possible but unproven—nor even weakly corroborated. Neutral models are a methodological tool appropriate to approach this problem.

REFERENCES

Hecht, M.K., and A. Hoffman. 1986. Why not neodarwinism? A critique of paleobiological challenges. *Oxford Surveys in Evolutionary Biology* 3:1–47.

Hempel, C.G., and P. Oppenheim. 1948. Studies in the logic of explanation. *Philosophy of Science* 15:135–175.

Hoffman, A. 1981. Stochastic versus deterministic approach to paleontology: The question of scaling or metaphysics? *Neues Jahrbuch für Geologie und Paläontologie, Abhandlungen* 162:80–96.

Hoffman, A. 1983. Paleobiology at the crossroads: A critique of some modern paleobiological research programs. In *Dimensions of Darwinism*, ed. M. Grene, pp. 241–271. Cambridge: Cambridge University Press.

Hoffman, A., and W.-E. Reif. On methodology of the biological sciences. From a biological perspective. Submitted for publication.

Mayr, E. 1982. *The Growth of Biological Thought*. Cambridge, Mass.: Harvard University Press.

Rosenberg, A. 1985. *The Structure of Biological Science*. Cambridge: Cambridge University Press.

Salthe, S.N. 1985. *Evolving Hierarchical Systems*. New York: Columbia University Press.

I
MOLECULAR AND GENETIC MODELS

1
Neutral Models in Molecular Evolution

JAMES F. CROW

In the spirit of W. C. Wimsatt's analysis in this volume, I shall treat the neutral theory of molecular evolution as a neutral or baseline model to be applied to data. Whether the model is correct is not the central issue for this purpose, but whether it is useful.

Ever since Darwin, the chief explanatory theory for understanding evolutionary adaptation has been natural selection. Differential survival and reproduction, acting on genetic variation, has been regarded as a sufficient mechanism. The seeming purposefulness of innumerable structures and processes is attributed to differential propagation of the fittest. Darwin made a respectable word out of teleology—or, to be verbally more fastidious, teleonomy.

The Darwinian theory is also heuristic. Time after time the assumption that a structure or process is there because it was favored by selection and therefore performed some useful function has paid off. A search for the function has led to a deeper understanding of both the function and the mechanism. As a guiding principle, natural selection has certainly stood the test of time and proven its heuristic value.

Of course, it is possible to go overboard and attribute every wrinkle to natural selection, as Gould and Lewontin (1979) have taken such delight in pointing out. My own favorite candidates for exuberant panselectionism are those geneticists and molecular biologists who assume that mutation rates are somehow controlled so that the species responds to an environmental change by increasing its mutation rate and thereby more quickly adapts to the new environment. I think these people overlook the very loose coupling between a suddenly increased mutation rate and the amount of selectable genetic variance, and also the difficulty of adjusting mutation rates by natural selection in a sexual population.

Natural selection occurs at many levels and in many guises: individual selection, within-family selection, sexual selection, kin selection, group selection, and species selection. Many alternatives exist to direct natural selection as explanations of evolutionary change. There is natural selection at lower levels, between cells (as in gametic competition, or malignancy), between chromosomes (as preferential inclusion of B chromosomes in the fertilizing nucleus of the corn pollen tube), and between homologous genes or linked groups (as in segregation distortion). There are additional mechanisms associated with transposable elements and retroviruses that can exert evolutionary influences (Dover and Flavell 1982; Shapiro 1983; Temin and Engels 1984). It is not always obvious whether such molecular mechanisms are best regarded as an extension of Darwinian selection to the subcellular level, or are better described by such terms as "molecular drive" (Dover 1982).

The molecular revolution has not removed natural selection from its central position in explaining adaptive changes. On the contrary, it has enriched the Darwinian theory by providing more details of how selection acts, particularly selection at the organismic level and below. When we look at evolution of the genetic material itself, however, we find that organismic selection appears to be less important, while other forces intrude and may become overriding. In particular, many changes at the DNA level have so little effect on survival and fertility of the host organism that their evolution is determined largely by chance. The driving forces then become the properties of the genetic material itself—mutation, gene conversion, transposition, gene duplication, and all the processes that have been ascribed to "selfish DNA."

The theory that most molecular evolution can best be accounted for by these forces, together with a stochastic treatment of the processes by which such changes are incorporated into the population, has been remarkably successful, both qualitatively and quantitatively. The person mainly responsible for the theory and its mathematical development is Motoo Kimura (1968, 1983), who has presented his arguments with force and conviction.

The neutral theory has not gone unchallenged, nor has it remained unchanged. It might be better to call it the theory of evolution by mutation and random drift, for it no longer asserts strict neutrality—only that, in most molecular evolution, selection is so weak that other forces predominate. To quote from Kimura's book:

> The neutral theory asserts that the great majority of evolutionary changes at the molecular level, as revealed by comparative studies of protein and DNA sequences, are caused not by Darwinian selection but by random drift of selectively neutral or nearly neutral mutants. The theory does not deny the role of natural selection in determining the course of adaptive evolution, but it assumes that only a minute fraction of DNA changes in evolution are adaptive in nature, while the great majority of phenotypically silent molecular substitutions exert no significant influence on survival and reproduction and drift randomly through the species (Kimura 1983, p. xi).

In its latest form the theory assumes that the majority of incorporated mutations are actually *very* slightly deleterious, a modification that owes a great deal to Ohta (1976).

I do not propose to discuss the various arguments for and against the mutation-random drift theory. They have been rehashed many times. I propose instead to discuss some ways in which the theory has been (1) predictive, (2) heuristic, (3) explanatory, and (4) useful. In addition, since it is possible to work out the consequences of random drift acting on point mutations and other DNA changes, the neutral theory provides a null hypothesis for testing other evolutionary factors, a neutral model of the kind Wimsatt and others in this volume have discussed.

PREDICTIVE VALUE OF THE NEUTRAL MODEL

The theory of molecular evolution by mutation and random drift led to Kimura's prediction that those amino acid or nucleotide changes that have the least consequence would evolve fastest. This was seen in the hemoglobin molecule, where those amino acids in contact with the important heme group change very slowly, whereas those on the outside of the molecule, where the constraints are fewer, change more rapidly. Likewise, the parts of the proinsulin molecule that are processed out as the functional insulin is formed evolve about six times as rapidly as the functional parts (Kimura 1983, p. 159 ff).

With the accumulation of data from DNA sequences the predictions were strikingly borne out. Synonymous changes in the genetic code (i.e., nucleotide changes that do not change the amino acid) evolve more rapidly than nonsynonymous ones. This is especially striking in those proteins that have been very slow-changing, such as histones. Histone 4 is perhaps the most conservative of all proteins; its amino acid substitution rate is only about 1% that of hemoglobins. Yet synonymous changes occur at about the same rate in histone 4 and hemoglobins (Kimura 1983, p. 173 ff).

Another neutral prediction is that the rate of molecular evolution is largely independent of the environment and of the rate of morphological evolution. This, too, is verified. Hemoglobin has evolved at about the same rate in fish as in land mammals, despite their totally different habitats. Perhaps the most striking examples are "living fossils"—species that have hardly changed in morphology over long geological times. The Port Jackson shark is a relict survivor of a Paleozoic species, yet its α and β hemoglobins differ from each other by as much as those of the human (Kimura 1983, p. 81).

Clearly, the neutral model has had useful predictive properties (or, in some instances, retrodictive). One might argue, with Fisher (1930, p. 41 ff), that the smaller the change caused by a mutation, the greater the probability that the change is beneficial. Thus, those changes whose effects are smallest would be more likely to be incorporated by natural selection, and this would be reflected in a more rapid evolution rate. So one could say that the same prediction might have been made from a selectionist theory.

One neutralist response to this argument is that although this prediction might have been made on selectionist grounds, it wasn't. A second response is this: The rate of evolutionary substitution of rare, favorable mutants is approximately $2N\mu P$, where N is the population number, μ is the rate of occurrence per generation of favorable mutations, and P is the probability that a mutant will ultimately be fixed. For a slightly favorable mutant, $P \approx 2s$, where s is the selective advantage of the mutant allele, as first shown by Haldane (1927). Making this substitution, the rate of amino acid or nucleotide replacement is $4N\mu\bar{s}$, where $\bar{s}$ is the average selective advantage of the favorable mutants. This says that the evolutionary rate is proportional to the mutation rate and to the selective advantage, as expected. But it also says that the rate is proportional to the species population number, which is contrary to fact. Invertebrates evolve molecularly no faster than vertebrates, and common herbivores change no faster than rare carnivores. This argues strongly against molecular evolution being mainly the incorporation of a succession of favorable mutations. For a more detailed discussion, see Kimura (1983).

One can contrive alternative selectionist arguments. For example, it is sometimes argued that favorable mutations occur only after an environmental change, and that the time spent waiting for such a change is not a function of population number. A more sophisticated version of this argument has been given by Gillespie (1984). As I said earlier, however, it is not my object in this article to review the arguments for and against the correctness of the neutral model, but to point out its predictive value, which these examples illustrate strikingly.

HEURISTIC VALUE OF THE NEUTRAL MODEL

The neutral model has had a revolutionary effect on population genetics, both theoretical and experimental. It has led, directly or indirectly, to investigation of such questions as the number of generations required for a neutral mutant to drift to fixation; the answer turns out to be surprisingly simple—four times the effective population number. The theory has also led to an emphasis on the number of gene substitutions that take place over a very long time period, which is not strongly dependent on population size, and the time required for an individual substitution to take place, which is. This and other earlier results are reviewed in Crow and Kimura (1970). The diffusion models used to solve neutral problems are equally useful for weak deterministic forces, such as selection or migration. For example, one can ask what is the total number of individuals affected by a mutant gene, whether it be beneficial, harmful, or neutral, during its lifetime in the population. Or one can ask for the total number of heterozygotes for a mutant gene from the time of its occurrence (Maruyama and Kimura 1971), a measure of importance because of the central role of variance in population genetics. It has led to a successful search for quantities that have invariance properties (Maruyama 1977; Nagylaki 1982). It has led to more rigorous theories of clines, following the pioneer work of Haldane and Fisher (Nagylaki 1978). Much of this work was initiated by Kimura and his associates, but many others have participated in a rapid increase of sophisticated theory. It would be incorrect to attribute all of this to the neutral model, but the neutral model has often provided the seedling onto which considerations of selection and other forces could be grafted.

Likewise, the neutral theory has stimulated a great deal of experimental work. Much of this has been done in an attempt to test the theory itself, or to prove or disprove it. This effort to provide definitive tests of the theory has not been a success. Yet, the existence of the neutral model has led to more systematic and relevant observations than would otherwise have been the case. Whatever the ultimate acceptance of the neutral theory, it will have played a major role in the guidance of a generation of experimental and theoretical population genetics.

EXPLANATORY VALUE OF THE NEUTRAL MODEL

Neutral models, like almost everything else in evolution, might be said to have had their start with Darwin, who in *The Origin of Species* referred to some traits as conferring no selective advantage or disadvantage. In molecular terms the first to propose neutrality for explanatory purposes were Freese (1962) and Sueoka (1962). They were puzzled by the fact that bacterial species with roughly the same amino acid content differed greatly in the amounts of the four nucleotides. Some bacteria are rich in A and T, 75% or more, while others are equally rich in G and C. To explain this, they suggested that in some species the mutation rate from AT to GC is higher than that from GC to AT, while in other species the situation is reversed.

Coming in advance of knowledge of the code or of its universality, this showed remarkable insight. Now, of course, we know that the code is highly redundant. If there is mutation pressure in favor of A and T, we should expect these bases to predominate strongly in synonymous positions, and at other positions to a lesser extent, depending on whether the changes are or are not selectively important.

Recent work by Osawa and his colleagues (Yamao et al. 1985; Muto et al. 1985) has shown that in *Mycoplasma capricolum* 75% of the bases are A and T and only 25% are G and C. The situation is still more extreme if we examine the third codon position where almost all changes from GC to AT are synonymous.

I am going to refer to RNA codons instead of DNA, so that T is replaced by U. Among 1,090 codons, 978 or 90% have A or U at the usually synonymous third position and only 112 have G or C. Furthermore, 34 of these 112 are AUG, the start codon, which cannot be changed without upsetting the whole protein-synthesizing system. So there are only 78 exceptions in 1,056 codons, or 7%, to having either A or U rather than G or C whenever there is a choice. There are known DNA polymerases that tend to make replication errors in the direction that favors mutation from GC to AT, so it makes sense to assume that something like this has happened in the history of *Mycoplasma*. It is also possible, though perhaps less likely, that there is some overall selective constraint on the AT:GC ratio that favors AT in this species, so that, whenever feasible, AT replaces GC.

The most remarkable of Osawa's results is the finding that *Mycoplasma* uses a different genetic code. The genetics world had one of its firmest dogmas upset a few years ago when it was discovered that mitochondria use a different code than the rest of the cell. Now, so do *Mycoplasma* and some of the ciliates. One would think that the code, whether or not it is the best, would be unchangeable, since changing the code would mean that all sorts of wrong amino acids would be incorporated into proteins with results that would surely be lethal. How can we explain these code changes?

I am indebted to Dr. Thomas Jukes, who with the late Jack L. King independently proposed the neutral hypothesis (King and Jukes 1969), for permission to present his clever solution to the *Mycoplasma* code dilemma.

Figure 1.1 shows the relevant portion of the messenger RNA code dictionary for "normal" organisms, for mitochondria, and for *Mycoplasma*. In mitochondria UGA has changed from a stop codon to one that codes for tryptophan. *Mycoplasma* has gone a step further, and UGG is missing.

Figure 1.2 shows a way in which this might possibly have come about, using a neutral, mutation-driven model. I have included both the codon in the messenger RNA and the anticodon in the transfer RNA, which pair according to standard Watson-Crick rules. Successive mutational changes are numbered.

```
        Standard            Mitochondria          Mycoplasma

   UAG .  UGG Tryp      UAG .  UGG Tryp      UAG .  UGG

   UAA .  UGA .         UAA .  UGA Tryp      UAA .  UGA Tryp
```

Figure 1.1. A portion of the standard mRNA code dictionary along with the same region for mitochondria and *Mycoplasma capricolum*. Stop codons are indicated by a period.

Notice that the first type of mutation, from codon UGA to UAA, might be expected to be neutral, since there is no change in function. In the absence of selection pressure, mutations in this direction would be expected to accumulate. The rate, of course, would be very slow, but we have billions of years at our disposal, since the divergence of *Mycoplasma* and other organisms from a common ancestral type must have been a very long time ago. Eventually, however, all (or essentially all) UGA codons would be replaced by UAA.

When this state is attained, then a mutation in the gene responsible for the transfer RNA changing the anticodon ACC to ACU would have no consequence, for there would be other transfer RNAs containing the ACC anticodon to carry on the translation. This change, numbered 2, from C to U, is in the direction favored by mutation. After a sufficient number of ACC anticodons have changed to ACU, a mutation in the UGG codon to UGA, change number 3, would do no harm. It would still be translated into tryptophan. It might even be beneficial, if by this time there were more ACU than ACC anticodons, by increasing the production of tryptophan. Eventually mutation pressure of types 2 and 3 would effectively eliminate UGG from the genome. Note that the mitochondrial code is equivalent to one of the intermediate steps in the process. In fact Osawa has noted that the mitochondrial code is intermediate between those of normal organisms and *Mycoplasma,* an item of interest in view of the suggestion that mitochondria are of extracellular origin (Margulis 1981).

I don't want to argue that Jukes's scheme is necessarily correct. But it has great explanatory value. It provides a reasonable mechanism, not calling for any implausible events, for a change in the code—something that only recently was regarded as essentially impossible.

This model has the additional virtue of being heuristic. It immediately suggests measuring the mutation rates in *Mycoplasma* to see if GC → AT changes predominate. It also suggests further study of the stop codons, since two successive stop signals are often required. Could one of them be UGA? Above all, it suggests looking for deviations from the standard code dictionary in other species.

```
        Codon        UAG .         UGG Tryp
        Anticodon    AUC       3  ACC
                                        ) 2
        Codon        UAA . ←——1—— UGA .   (Tryp)
        Anticodon    AUU          ACU
```

Figure 1.2. Diagram showing how three successive sets of mutations, over a very long time period, can change the standard code to that for *Mycoplasma*. Anticodons are in italics and those bases that have changed are underlined.

MOLECULAR EVOLUTION

This is, I think, a particularly good example of the explanatory power of the neutral model. In this case the model also suggests experiments that are likely to be decisive, since bacteria lend themselves to experiments with great statistical resolving power.

USEFULNESS OF THE NEUTRAL MODEL

Many examples could be given of the practical utility of neutral gene models when studying other problems. Many laboratory strains are maintained as small populations where random drift predominates, even for characters that are ordinarily subject to natural selection. The extent to which lines deviate can be computed from the neutral model and this provides useful information to the experimenter. A second use is in designing breeding programs for rare species kept in captivity. It may be desirable, on the one hand, to try to adapt these to permanent zoo life or, on the other, to try to keep them as near as possible to their original state for possible future release. In either case, neutral models provide useful information. These and other problems have been discussed, along with a neutral theory of phenotypic evolution, by Lynch and Hill (1985).

I should like to use an example from my own work with Dr. Kenichi Aoki. It is well known that altruistic traits can develop by the principle of kin selection. It is also of interest to ask whether, in partially isolated groups, there is enough local differentiation for a similar principle to operate. In a subdivided population, members of a group are related to one another. Could traits that make individuals altruistic toward members of whatever group they belong to evolve by principles analogous to kin selection? This would obviate any need for kin recognition.

We were able to show that, in a subdivided population with a small amount of migration between groups, the coefficient of relationship within a group is $r = 2G_{ST}/(1 + G_{ST})$, where G_{ST} is Nei's (1975) measure of group divergence. From this we were able to recover Hamilton's cost-benefit condition, $r > c/b$, for a quantitative trait to increase, where c is the cost in reduced fitness per unit increase in the trait and $b - c$ is the expected increase in group fitness per unit increase in the average value of the trait (Crow and Aoki 1982).

The question is this: Can we devise a useful measure of the degree of population subdivision that will let us estimate the level of c/b that would be favored by natural selection acting both between and within groups?

We should like to make use of molecular data because of their ease and reproducibility of measurement, and for the abundance of information which they provide. Molecular polymorphisms are near enough to neutrality that a neutral model can be used. That is to say, we want to use neutral or nearly neutral molecular changes to assess the degree of population subdivision and from that infer what would happen to selected traits with such a population structure.

We define H_0 as the probability that two alleles, drawn at random from the same subpopulation, are not identical, and $\bar{H}$ is the probability for two alleles drawn at random from the entire population. Then the degree of subdivision is conveniently measured by

$$(1.1) \quad G_{ST} = (\bar{H} - H_0)/\bar{H}$$

The relevant parameters are μ, the mutation rate; K, the number of possible alleles at the locus in question; N, the size of the group; m, the rate at which migrants are exchanged between a group and the other groups per generation; and n, the number of groups. We can write recurrence relations between the value of H_0 and $\bar{H}$ for successive generations and, by equating the values in successive generations, solve for equilibrium values. The equilibrium solutions are of the form

$$(1.2) \quad \hat{\bar{H}} = f(\mu,N,m,n,K)$$
$$\hat{H}_0 = g(\mu,N,m,n,K)$$

that is to say, they are functions of all five parameters.

For explicit solutions and further details, see Crow and Aoki (1984). Not only do these solutions involve all five parameters, most of which are difficult or impossible to measure, but the rate of approach to the equilibria is very slow, being governed by the mutation rate and the total population size. So the formulae are not of much practical use.

Yet, when Aoki and I looked at G_{ST} we had a pleasant surprise. Assuming that $\mu << m$, $1/N << 1$ and neglecting terms of order μ^2, μm, and μ/N, we find the equilibrium value

$$(1.3) \quad \hat{G}_{ST} \approx 1/(4M\alpha + 1)$$

where

$$M = Nm \quad \text{and} \quad \alpha = [n/(n-1)]^2$$

Thus, unless the number of groups is very small, the equilibrium value, $\hat{G}_{ST}$, depends on only one parameter, M, the absolute number of migrants per generation. This quantity is comparatively easy to assess, but most important is the fact that the difficult quantities, μ, n, and K have dropped out. Furthermore, although H_0 and H approach equilibrium very slowly, G_{ST} approaches it fairly rapidly, being governed by m and $1/N$.

G_{ST}, then, is readily calculated from molecular data as a function of two readily measured quantities, H_0 and $\bar{H}$ (neither of which is of much practical interest), approaches equilibrium rapidly, depends mainly on only one parameter—the absolute migration rate—and is directly related to Hamilton's cost/benefit ratio.

The data available on Japanese monkeys suggest that the population structure is such that, if present values are representative of the past, traits whose cost/benefit ratios are less than 0.10–0.15 would be selected. Many more data are needed before we can know whether these values are typical. The assumption here is that the trait is such that an individual behaves in this manner in whatever group it grows up in, regardless of whether its parents were migrants. I think this is probably a reasonable model. Of course this whole discussion assumes that such traits exist and have a heritability greater than zero.

To summarize, we have used a neutral model as a way of developing a theory that can be used to assess the degree of population subdivision and to see what such a structure implies for the evolution by natural selection of traits that are harmful to the individual but benefit the group.

IN PRAISE OF VERY WEAKLY SELECTED GENES

Whatever one's view of the neutral theory in its strict form, there can be little room for doubt that it, together with molecular developments, has shown that there are many alleles whose selective differences are very small.

Most quantitative traits are polygenic and the maximum fitness is associated with an intermediate value of the trait. Usually the fitness is a concave function, such that near the optimum small displacements of the trait value cause very little reduction in fitness but large displacements cause a large decrease. Haldane and Wright both considered a model in which fitness decreases with the square of the distance from the optimum.

It seems to me that molecular discoveries and the evidence for very weak selection add to the validity of Darwinian gradualism. I can see a number of evolutionary advantages in having very weakly selected alleles that are part of a polygenic system. The polygenic ideal of a large number of independent, additive genes provides in some ways the maximum opportunity for Mendelism to be effective. Here are what I regard as a set of reasonable, although hardly original, conclusions:

1. The smaller the effect of an allele, the more nearly additive it is. Within a locus there is little dominance; between loci there is little epistasis. This is both an empirical observation and a theoretical expectation (Mukai et al. 1972; Kacser and Burns 1981).

2. The more nearly additive the genes affecting a trait are, the more responsive the trait is to selection. Selection in a Mendelian population acts on the additive component of the genetic variance, not the total genetic variance.

3. The larger the number of genes determining a trait and the smaller the effect of each, the greater is the opportunity for fine tuning. There is essentially a continuum of possible phenotypic values, giving the opportunity to adjust with a precision tool rather than a sledge hammer.

4. Most important of all, multiple factors with approximately additive effects and more or less interchangeable functions provide a maximum of potential variance with a minimum of standing variance. Numerous experiments have shown that a few generations of selection can produce a population in which the average individual far exceeds the extremes of the founding population. This means that a species does not have to carry around a large load of deviant individuals in order to be responsive to changes in the environment. It can quickly respond to an environmental change that alters the position of the optimum phenotype.

In addition to their obvious advantages in fitness evolution, multiple, additive, independently inherited genes constitute another null model, which has traditionally served as a baseline from which deviations and interactions, often of great interest, can be discovered and measured.

Of course the real world isn't so simple. Nature doesn't necessarily operate in the way that is most appealing to this particular view. There are dominance and epistasis. If the chromosome number is small, linkage may play a role. Characters are correlated; selection for one trait may cause maladaptive changes in another because of pleiotropic or allometric relations. To some extent pleiotropy and other causes of genetic correlations *do* impede selection experiments. Especially if there is long-continued selection in a new direction, unwanted by-products appear. In one experiment, mice selected for large size became very large, but were sluggish, while those selected for small size had low fertility. Yet, such correlations can usually be broken. If every change in liver function affected the brain, selection for improved livers would lead to mental retardation. Often, I suspect most of the time, if there is a gene that produces two effects there is another, or a modifier, that affects one or the other. Perhaps those species that have been most successful in evolution, and are therefore still around, are those in which a certain amount of independence and continuity was present in the genetic system of their ancestors.

The question is a complicated one and far from solved, as the discussion by Stuart Kauffman (Chapter 3) has shown. There is strong evidence for some sort of genetic coordinating principle; a hybrid between a greyhound and a chihuahua doesn't have an upper jaw like a greyhound and a lower jaw like a chihuahua, even in the F_2 generation. Yet it has been possible to select for an undershot jaw in bulldogs. Sewall Wright made a pioneer study determining how much of the variance in the distal part of the forelimb was due to overall size factors, how much to limb factors, and how much to special factors. All made significant contributions. For a summary, see Wright (1968, p. 330 ff). The developmental system seems capable of being affected by selection for both general factors and special ones at the same time.

Sewall Wright has long argued that interactions are a serious impediment to mass selection. In order for a population to change from having one coadapted complex to another, better one, it may be necessary to pass through allele frequencies that render the average fitness less. This cannot happen by mass selection of independent loci.

In this case, Wright has postulated that a population structure that permits some random gene-frequency drift in subpopulations may permit a local population to acquire such a favorable set of allele frequencies. These alleles can then be exported to the entire population by one-way migration and intergroup selection. For recent reviews, see Wright (1980, 1982).

Although Wright's and Kimura's models have quite different purposes, they both assume that there are numerous alleles that are so weakly selected that their fate is largely determined by random processes. Wright, however, is thinking of drift in small local populations, whereas Kimura is thinking of the whole species. Hence Kimura's near-neutrality is much closer to absolute neutrality than is Wright's.

It is clear that much of evolution depends on very small changes, and molecular genetics has provided the detailed evidence for the existence of many gene differences that are, at most, very weakly selected. In my view, this adds further strength to the Darwinian theory of evolutionary gradualism.

ACKNOWLEDGMENT

This is paper number 2823 from the Laboratory of Genetics, University of Wisconsin.

REFERENCES

Crow, J.F., and K. Aoki. 1982. Group selection for a polygenic behavioral trait: A differential proliferation model. *Proceedings of the National Academy of Sciences, USA* 79:2628–2631.
Crow, J.F., and K. Aoki. 1984. Group selection for a polygenic behavioral trait: Estimating the degree of population subdivision. *Proceedings of the National Academy of Sciences, USA* 81:6073–6077.
Crow, J.F., and M. Kimura. 1970. *An Introduction to Population Genetics Theory*. New York: Harper and Row. Reprinted 1976. Minneapolis: Burgess Publishing Company.
Dover, G.A. 1982. Molecular drive: A cohesive mode of species formation. *Nature* 299:111–117.
Dover, G.A., and R.B. Flavell, eds. 1982. *Genome Evolution*. New York: Academic Press.
Fisher, R.A. 1930. *The Genetical Theory of Natural Selection*. Oxford: Oxford University Press. 2nd ed., 1958. New York: Dover Publications.
Freese, E. 1962. On the evolution of base composition of DNA. *Journal of Theoretical Biology* 3:82–101.
Gillespie, J.H. 1984. Molecular evolution over the mutational landscape. *Evolution* 38:1116–1129.
Gould, S.J., and R.C. Lewontin. 1979. The spandrels of San Marco and the Panglossian paradigm: A critique of the adaptationist programme. *Proceedings of the Royal Society of London (B)* 205:581–598.
Haldane, J.B.S. 1927. A mathematical theory of natural and artificial selection. Part V. Selection and mutation. *Proceedings of the Cambridge Philosophical Society* 23:838–844.
Kacser, H., and J.A. Burns. 1981. The molecular basis of dominance. *Genetics* 97:639–666.
Kimura, M. 1968. Evolutionary rate at the molecular level. *Nature* 217:624–626.
Kimura, M. 1983. *The Neutral Theory of Molecular Evolution*. Cambridge: Cambridge University Press.
King, J.L., and T.H. Jukes. 1969. Non-Darwinian evolution. *Science* 164:788–798.
Lynch, M., and W. Hill. 1985. The neutral theory of phenotypic evolution. Submitted for publication.
Margulis, L. 1981. *Symbiosis in Cell Evolution*. San Francisco: Freeman Publishing Company.
Maruyama, T. 1977. *Stochastic Problems in Population Genetics*. Lecture Notes in Biomathematics 17. Berlin: Springer-Verlag.
Maruyama, T., and M. Kimura. 1971. Some methods for treating continuous stochastic processes in population genetics. *Japanese Journal of Genetics* 46:407–410.
Mukai, T., S.I. Chigusa, L.E. Mettler, and J.F. Crow. 1972. Mutation rate and dominance of genes affecting viability in *Drosophila melanogaster*. *Genetics* 72:335–355.
Muto, A., F. Yamao, Y. Kawauchi, and S. Osawa. 1985. Codon usage in *Mycoplasma capricolum*. *Proceedings of the Japan Academy (B)* 61:12–15.

Nagylaki, T. 1978. Random genetic drift in a cline. *Proceedings of the National Academy of Sciences, USA* 75:423–426.

Nagylaki, T. 1982. Geographical invariance in population genetics. *Journal of Theoretical Biology* 99:159–172.

Nei, M. 1975. *Molecular Population Genetics and Evolution*. Amsterdam: North Holland Publishing Company.

Ohta, T. 1976. Role of very slightly deleterious mutations in molecular evolution and polymorphism. *Nature* 252:351–354.

Shapiro, J.A., ed. 1983. *Mobile Genetic Elements*. New York: Academic Press.

Sueoka, N. 1962. On the genetic basis of variation and heterogeneity of DNA base composition. *Proceedings of the National Academy of Sciences, USA* 48:582–592.

Temin, H.M., and W.R. Engels. 1984. Movable genetic elements and evolution. In *Trends in Evolution*, ed. J.W. Pollard, pp. 173–201. New York: John Wiley & Sons.

Wright, S. 1968. *Evolution and the Genetics of Populations. Vol. 1. Genetic and Biometric Foundations*. Chicago: University of Chicago Press.

Wright, S. 1980. Genic and organismic selection. *Evolution* 34:825–843.

Wright, S. 1982. Character change, speciation, and the higher taxa. *Evolution* 36:427–443.

Yamao, F., A. Muto, Y. Kawauchi, M. Iwami, S. Iwagami, Y. Azumi, and S. Osawa. 1985. UGA is read as tryptophan in *Mycoplasma capricolum*. *Proceedings of the National Academy of Sciences, USA* 82:2306–2309.

2
False Models as Means to Truer Theories

WILLIAM C. WIMSATT

Many philosophers of science today argue that scientific realism is false. They often mean different things by this claim, but most would agree in arguing against the view that scientific theories give, aim to give, approximate, or approach asymptotically to give a true description of the world. All theories, even the best, make idealizations or other false assumptions that fail as correct descriptions of the world. The opponents of scientific realism argue that the success or failure of these theories must therefore be independent of, or at least not solely a product of, how well they describe the world. If theories have this problematic status, models must be even worse, for models are usually assumed to be mere heuristic tools to be used in making predictions or in aiding the search for explanations, and which only occasionally are promoted to the status of theories when they are found not to be as false as assumed.

Although this rough caricature of philosophical opinion may have some truth behind it, these or similar views have led most writers to ignore the role that false models can have in improving our descriptions and explanations of the world. (Cartwright's [1983] excellent studies of the use and functions of models are a striking exception here.) Although normally treated as a handicap, the falsity of scientific models is in fact often essential to this role. I will not discuss the larger issue of scientific realism here: The way in which most philosophers have formulated (or misformulated) that problem renders it largely irrelevant to the concerns of most scientists who would call themselves realists. Philosophers attack a realism which is "global" and metaphysical. Most scientists use and would defend a more modest (or "local") realism, and would do so on heuristic rather than on metaphysical grounds.

By local realism I mean something like the following: On certain grounds (usually, for example, that the existence of an entity or property is known, derivable, or detectable through a variety of independent means; see Wimsatt 1981a), scientists would argue that an entity or property is real, and they cannot imagine plausible or possible theoretical changes that could undercut this conclusion. Furthermore, they might argue that their experimental and problem-solving approaches require them to presuppose the existence of that entity, property, or phenomenon—a heuristic argument. I suspect that many philosophical opponents of scientific realism could accept this kind of local and heuristic realism. I think that it is the kind of realism most worth defending, though I think that it may also give handles for defending a more ambitious kind of scientific realism. In any case, there is much more of use to be found in the topic of false models, to which I will now turn.

Neutral models will be discussed mostly indirectly in this chapter. After an attempt at characterizing what a neutral model is, I will show that this idea relates more generally to the use of false models, and then discuss a variety of ways in which false models are used to get to what we, at least provisionally, regard as the truth. Many of the chapters in this volume turn out to use their models in one or more of these ways. Thus, the focus of my chapter may be conceived as generalizing beyond the case of neutral models to consider the variety of uses of models, in spite of, or even because of, the fact that they are assumed to be false, to get at the truth or our best current approximations to it.

EVEN THE BEST MODELS HAVE "BIASES"

The term "neutral model" is a misnomer if it is taken to suggest that a model is free of biases such as might be induced by acceptance of a given hypothesis (such as that the patterns to be found among organisms are products of selection). Any model must make some assumptions and simplifications, many of which are problematic, so the best working hypothesis would be that there are no bias-free models in science.

This observation has a parallel in the question, "What variables must be controlled for in an experimental design?" There are no general specifications for what variables should be controlled, since (1) what variables should be controlled or factored out through appropriate attempts to isolate the system, (2) what variables should be measured, (3) what errors are acceptable, and (4) how the experiment should be designed, are all functions of the purpose of the experiment. Similarly, (1) what models are acceptable, (2) what data are relevant to them, and (3) what counts as a "sufficiently close fit" between model and data are functions of the purposes for which the models and data are employed. (As one referee for this chapter pointed out, many people who would applaud the use of appropriate controls and isolations in an experimental design inconsistently turn on mathematical or causal models of the system and criticize them for doing the same thing! This activity is desirable and necessary in either case.)

Any model implicitly or explicitly makes simplifications, ignores variables, and simplifies or ignores interactions among the variables in the models and among possibly relevant variables not included in the model. These omitted and simplified variables and interactions are sources of bias in cases where they are important. Sometimes certain kinds of variables are systematically ignored. Thus, in reductionistic modeling, where one seeks to understand the behavior of a system in terms of the interactions of its parts, a variety of model-building strategies and heuristics lead us to ignore features of the environment of the system being studied (Wimsatt 1980b, pp. 231–235). These environmental variables may be left out of the model completely, or if included, treated as constant in space or in time, or as varying in some particularly simple way, such as in a linear or random fashion. In testing the models, with the focus on the interrelations among internal factors, environmental variables may be simply ignored (Wimsatt 1980b), or treated in some aggregate way to simplify their analysis (Taylor 1985; Wimsatt 1985).

A model may be bias-free in one case (where its variables and parameters refer, at least approximately correctly, to causal factors in nature, and where its "accessory conditions" [Taylor 1985] are satisfied) but biased for other cases it is applied to because it worked well in the first case. (Taylor [1985] discusses examples of this for ecological models in greater detail.) Even where it is recognized that a model must be changed, there may be biases (1) in how or where it is changed, or (2) in the criteria accepted for recognizing that it must be changed.

An example of the first type is found in reductionistic modeling: Whereas both the description and interaction of parts internal to and of variables external to the system are usually oversimplified, there will often be a bias toward increasing the internal realism of the model in cases where the failure of fit of the model with the data is due to unrealistic assumptions about the environment (Wimsatt 1980b).

A potential example of the second kind is provided by Williams's (1966, p. 17) "principle of parsimony," in which he recommends "recognizing adaptation at no higher level than is required" (presumably by the data). If this recommendation is taken as an invitation to find *some* set of parameter values for the simple model for which it fits the data, then one may be engaged in a "curve-fitting" exercise which may hide the need for a higher-level or more complex model, unless one takes the pains to determine whether the parameter values for which fit is achieved are actually found in nature. This is seldom done, at least in the kind of optimization modeling frequently found in evolutionary biology.

THE CONCEPT OF A "NEUTRAL MODEL"

In evolutionary biology and ecology, a "neutral model" usually means a model without selection. Thus, Raup (Chapter 6) in his work with Gould, Schopf, and Simberloff considered "random phylogenies," phylogenetic descent trees in which originations and extinctions are determined by random variables. These artificial phylogenies in many respects resemble those found in nature. They thus reasoned that the similarities between the artificial and natural phylogenies were not products of selection processes operating at that level. Their model did not, as they pointed out, rule out selectionist explanations for speciations and extinctions at a lower (e.g., intra- or interpopulational) level. Similarly, the work of Crow (Chapter 1) and others on "neutral mutation" theories modeled and evaluated patterns of molecular variability and change on the assumption that selection forces on these variants could be ignored. Similarities between their predicted patterns and what was found in nature led to various versions of the hypothesis that the evolution of various systems or of various kinds of traits was driven not by selection, but by various forms of "genetic drift." Finally, Kauffman's work (1985 and Chapter 3) on "generic constraints" in development identified features which, because they were nearly universal properties of his randomly constructed model genetic control networks, are taken to provide "baselines" for the properties of systems of which selection acts. He argues that these "generic properties" will usually survive *in spite of* selection rather than because of it. Here Kauffman is using a comparison of models with and without selection to argue that selection is not important.

One must not assume that if the data fit the "neutral" model then the excluded variables are unimportant. The researchers of Raup, Crow, and Kauffman each suggest that selection may be unimportant in some way in producing the phenomena being modeled, but they do not rule it out entirely. Thus, Raup's "random phylogenies" do not exclude selection as a causal agent in producing individual extinctions (e.g., through overspecialization to a temporally unstable niche) or speciations (e.g., through directional selection in different directions producing isolating mechanisms between two different geographically isolated subpopulations of the same species), because his model simply does not address those questions.

Similarly, the work of Crow is consistent with the hypothesis that the very neutrality of most mutations is a product of selection. This could occur through selection for developmental canalization of macroscopic morphological traits, selection for redundancy of design through duplicated DNA, codon synonymy, or redundancy at the level of parallel synthetic paths in metabolic pathways, or at other higher levels. It could also represent selection for an architecture for the program of gene expression to preserve relatively high heritability of fitness or other phenotypic traits in the face of random sexual recombination, and uses any of the preceding mechanisms or others yet to be discovered to accomplish this.

Finally, the ubiquity of Kauffman's "generic properties" among genetic control networks of a certain type (which he takes as a constraint on his simulations) does not rule out the possibility that that type may itself be a result of prior selection processes. Thus, Kauffman's genetic control networks have an average of two inputs and outputs per control element. But he chose this value of these parameters in part because his own prior work (Kauffman 1969) showed that such networks had a shorter mean cycle time than networks containing more or fewer inputs and outputs, a feature which he argues is advantageous.

I have elaborated elsewhere a model of developmental processes (Wimsatt 1986) which presents an alternative hypothesis for the explanation of developmental constraints that seems likely in many such cases, and in even more cases where the feature in question is taxonomically very widely distributed, but not absolutely universal. Such broad or "generic" universality of some traits could be produced by strong stabilizing selection due to the dependence of a wide variety of other phenotypic features on them. This view is not new, but to my knowledge, no one has attempted to model its consequences, with the exception of a model proposed by Arthur (1982, 1984) to explain features of macroevolution. (Either Arthur's model or mine gives a plausible explanation, for example, for the near-universality of the genetic code. See, e.g., Kauffman 1985).

Jeffrey Schank and I have recently attempted to test my model by doing simulations on networks like those of Kauffman (1985), where connections do not contribute equally to fitness (as in Kauffman's selection simulations) but the contribution to fitness of a connection is a function of the number of nodes that are affected by that connection. We have tried three different fitness functions as measures of the topological properties of the connection and of the number of nodes that are "downstream" of it, with similar results in all cases. The results are quite striking. Our networks, like Kauffman's, show a decay in number of "good" connections under mutation and selection. This effect, which increases in strength with the size of the network, is what leads him to doubt that selection can maintain a network structure of any substantial size against mutational degradation. But if the number of nodes accessible to a given gene through its various connections is taken as a measure of "generative entrenchment" (Wimsatt 1986), then our results show that the genes (or connections of genes) that have been lost in this mutational decay are those with low degrees of generative entrenchment, and that virtually all of the genes (and connections) that are significantly generatively entrenched are still there after, 1,000 to 5,000 generations!

Thus, in effect, Kauffman's models confirm my theory, and the two theories need to be regarded as complementary rather than contradictory. We are doing a number of simulations to investigate the robustness of this phenomenon for different mutation rates, population sizes, numbers of generations, genome sizes, and connection densities. In the not too far distant future, it should be possible to do it for more realistic biochemical networks. We are now preparing some of these results for publication.

The models of Raup, Crow, and Kauffman use and rule out selection factors in different circumstances and in different ways, but they have two things in common. In each, the "neutral" model either does not include selection operating on the postulated variants, or (in Kauffman's case) supports arguments that selection is not efficacious under the circumstances considered. And in each, the "neutral" model is treated as specifying a "baseline" pattern with which natural phenomena and data are to be compared, in order to determine whether selection is required (if the natural phenomena do not fit the "neutral" pattern) or not (if they do fit). These "neutral models" are treated as "null hypotheses" (a term frequently found in the literature on ecological models, and of course in statistics; see Chapter 8), which are to be rejected only if the natural phenomena deviate sufficiently from those predicted by the model.

These cases suggest that we characterize a *"neutral model" as a "baseline model that makes assumptions that are often assumed to be false for the explicit purpose of evaluating the efficacy of variables that are not included in the model.* This leaves out selection, and thus perhaps falls short of a more complete characterization of neutral models in evolutionary biology, but this more general characterization makes more explicit the connection with hypothesis testing in statistics and allows us to bring in features of the use of models from other areas of biology to focus on the heuristic advantages of their use. The characterization of "neutral models" in this way leads naturally to the more general question of *when, how, and under what conditions models that are known or believed to be false can be used to get new or better information about the processes being modeled.* If "neutral" models are useful in biology, this has less to do with their "neutrality" than with more general features of the use of models in science.

The fit of data with a "neutral" model or "null hypothesis" usually establishes that omitted variables do not act in a way specific to the models under comparison, not that they do not act at all. This is consonant with my earlier claim that the adequacy of models is highly context-dependent, and that their adequacy for some purposes does not guarantee their adequacy in general. However, the use of these models as "templates" can focus attention specifically on where the models deviate from reality, leading to estimations of the magnitudes of the unincluded variables, or to the hypothesis of more detailed mechanisms of how and under what conditions these variables act and are important. This is a pattern of inference that is both common and important, and deserves closer scrutiny. The variety of ways in which this is done will be the primary focus of the remainder of the chapter.

AN OVERVIEW OF THE USES OF FALSE MODELS

The simple observation that most models are oversimplified, approximate, incomplete, and in other ways false gives little reason for using them. Their widespread use suggests that there must be other reasons. It is not enough to say (e.g., Simon 1981) that we cannot deal with the complexities of the real world, so simple models are all that we can work with, for unless they could help us do something in the task of investigating natural phenomena, there would be no reason for choosing model building over astrology or mystic revelation as a source of knowledge of the natural world.

Nor does the instrumentalist suggestion that we use them because they are effective tools rather than realistic descriptions of nature give us much help, for it presupposes what we want to understand—i.e., *how* false models can be effective tools in making predictions and generating explanations. I want to suggest a variety of ways in which false models can (1) lead to the detection and estimation of other relevant variables, (2) help to answer questions about more realistic models, (3) lead us to consider other models as ways of asking new questions about the models we already have, and (4) (in evolutionary or other historical contexts) determine the efficacy of forces that may not be present in the system under investigation but that may have had a role in producing the form that it has.

Before discussing the way in which false models can help us to find better ones, it is useful to have a classification of the ways in which a model can be false. By a model in this context, I mean one of two alternatives: (1) a mathematical model—an equation or set of equations together with the interpretations necessary to apply them in a given context, or (2) a causal model or proposed mechanism through which a phenomenon or set of phenomena is to be explained. Often one will have both, but the following comments apply roughly equally to either.

The following are ways in which a model can be false. They are ordered roughly in terms of increasing seriousness (except for 6 and 7):

1. A model may be of only very *local applicability*. In this case the model is false only if it is more broadly applied.
2. A model may be an *idealization* whose conditions of applicability are never

found in nature (e.g., point masses, the uses of continuous variables for population sizes, etc.) but which has a range of cases to which it may be more or less accurately applied as an approximation.
3. A model may be *incomplete,* leaving out one or more causally relevant variables. (Here it is assumed that the variables that are included are causally relevant, and are so in at least roughly the manner described.)
4. The incompleteness of the model may lead to a *misdescription of the interactions* of the variables that are included, producing apparent interactions where there are none ("spurious" correlations) or apparent independence where there are interactions—as in the spurious "context independence" produced by biases in reductionistic research strategies. Taylor (1985) analyzes the first kind of case for mathematical models in ecology, but most of his conclusions are generalizable to other contexts. In these cases, it is assumed that the variables identified in the models are at least approximately correctly described.
5. A model may give a *totally wrong-headed* picture of nature. Not only are the interactions wrong, but a significant number of the entities and/or their properties do not exist.
6. A closely related case is that in which a model is purely *"phenomenological."* That is, it is derived solely to give descriptions and/or predictions of phenomena without making any claims as to whether the variables in the model exist, e.g., the viral equation of state (a Taylor expansion of the ideal gas law in terms of T or V), automata theory (Turing machines) as a description of neural processing, or linear models as curve-fitting predictors for extrapolating trends.
7. A model may simply *fail to describe or predict the data* correctly. This involves just the basic recognition that it is false, and is consistent with any of the preceding states of affairs. But sometimes this may be all that is known.

Most of the cases to be discussed in this chapter represent errors of types 2 through 5, and the productive uses of false models would seem to be limited to cases of types 1 through 4 and 6. It would seem that the only context in which case 5 could be useful is where case 6 also applies, and often models that are regarded as seriously incorrect are kept as heuristic curve-fitting devices. There is no hard and fast distinction between phenomenological and nonphenomenological models, and the distinction between the two often appears to depend on context (see Cartwright 1983). Thus, the Haldane mapping function can be derived rigorously from first principles, but as such it makes unrealistic assumptions about the mechanisms of recombination. It is sometimes treated as an idealization (case 2) and sometimes as a phenomenological predictive equation (case 6). Finally, a model may make false or unrealistic assumptions about lower-level mechanisms, but still produce good results in predicting phenomena at the upper level. In this case, the upper-level relationships either may be robust (see Levins 1966; Wimsatt 1980a, 1981a) or may be quite fragile results which are secured through obvious or unintended "curve fitting," as with many optimization models in evolutionary biology and economics. The former is a productive use of falsehood (item 10 below), and the latter is an often unproductive instance of case 6 above.

It may seem paradoxical to claim that the falseness of a model may be essential to its role in producing better models. Isn't it always better to have a true model than a false one? Naturally, it is, but this is never a choice that we are given, and it is a choice which only philosophers could delight in imagining. Will any false model provide a road to the truth? Here the answer is just as obviously an emphatic "no!" Some models are so wrong, or their incorrectnesses so difficult to analyze, that we are better off looking elsewhere. Cases 5 and 7 above represent models with little useful purchase, and "curve-fitting" phenomenological models (case 6) would be relatively rarely useful for the kinds of error-correcting activity I propose. The most productive kinds of falsity for a model are cases 2 or 3 above, though cases of types 1 and 4 should sometimes produce useful insights. *The primary virtue a model must have if we are to learn from its failures is that it, and the experimental and heuristic tools we have available for analyzing it, are structured in such a way that we can localize its errors and attribute them to some parts, aspects, assumptions, or subcomponents of the model.* If we can do this, then "piecemeal engineering" can improve the model by modifying its offending parts.

There is a mythology among philosophers of science (the so-called "Quine-Duhem thesis"; see Glymour 1980) that this cannot be done, that a theory or model meets its experimental tests wholesale, and must be taken or rejected as a whole. Not only science, but also technology and evolution, would be impossible if this were true in this and in logically similar cases. That this thesis is false is demonstrated daily by scientists in their labs and studies, who modify experimental designs, models, and theories piecemeal; by electrical engineers, who localize faults in integrated circuits; and by auto mechanics and pathologists, who diagnose what is wrong in specific parts of our artifacts and our natural machines, and correct them. Glymour (1980) and Wimsatt (1981a) give general analyses of how this is done and descriptions of the revised view of our scientific methodology that results. The case for evolutionary processes is exactly analogous. Lewontin (1978) argues that without the "quasi-independence" of traits (which allows us to select for a given trait without changing a large number of other traits simultaneously) any mutation would be a (not-so-hopeful) monster, and evolution as we know it—a process of small piecemeal modifications—would be impossible (see also Wimsatt 1981b). In all of these cases, then, "piecemeal engineering" is both possible and necessary.

The following is a list of functions that false models may serve in the search for better ones:

1. An oversimplified model may act as a starting point in a series of models of increasing complexity and realism.
2. A known incorrect but otherwise suggestive model may undercut the too ready acceptance of a preferred hypothesis by suggesting new alternative lines for the explanation of the phenomena.
3. An incorrect model may suggest new predictive tests or new refinements

of an established model, or highlight specific features of it as particularly important.
4. An incomplete model may be used as a template that captures larger or otherwise more obvious effects, which can be "factored out" to detect phenomena that would otherwise be masked or be too small to be seen.
5. An incomplete model may be used as a template for estimating the magnitude of parameters that are not included in the model.
6. An oversimplified model may provide a simpler arena for answering questions about properties of more complex models that also appear in this simpler case, and answers derived here can sometimes be extended to cover the more complex models.
7. An incorrect simpler model can be used as a reference standard to evaluate causal claims about the effects of variables left out of it but included in more complete models, or in different competing models to determine how these models fare if these variables are left out.
8. Two false models may be used to define the extremes of a continuum of cases in which the real case is presumed to lie, but for which the more realistic intermediate models are too complex to analyze, or too special in their application to be of any general interest, or for which the information available is too incomplete to guide their construction or to determine a choice between them. In defining these extremes, the "limiting" models specify a property of which the real case is supposed to have an intermediate value.
9. A false model may suggest the form of a phenomenological relationship between the variables (a specific mathematical functional relationship which gives a "best fit" to the data but which is not derived from an underlying mechanical model). This "phenomenological law" gives a way of describing the data and (through interpolation or extrapolation) making new predictions, but it also, when its form is conditioned by an underlying model, may suggest a related mechanical model capable of explaining it.
10. A family of models of the same phenomenon, each one of which makes a variety of false assumptions, may be used in a variety of ways: (a) to look for results that are true in all of the models and therefore presumably independent of the various specific assumptions that vary from model to model (Levins's [1966] "robust theorems") and thus are more likely trustworthy or "true"; (b) in the same manner, to determine assumptions that are irrelevant to a given conclusion; and (c) where a result is true in some models and false in others, to determine which assumptions or conditions a given result depends upon (see Levins 1966, 1968; Wimsatt 1980a, 1981a).
11. A model that is incorrect by being incomplete may serve as a limiting case to test the adequacy of new, more complex models. (If the model is correct under special conditions, even if these are seldom or never found in nature, it may nonetheless be an adequacy condition or desideratum of newer models that they reduce to this model when appropriate limits are taken.)
12. Where optimization or adaptive design arguments are involved, an eval-

uation of systems or behaviors which are not found in nature, but which are conceivable alternatives to existing systems, can provide explanations for the features of those systems which are found.

The core of this chapter, which considers the development of the "chromosomal mechanics" of the Morgan school, is intended to illustrate the first point. Points 2 and 3 are not discussed here, but all of the remaining ones are. In the next sections, I will illustrate these functions by reanalyzing a debate between members of the Morgan school and W. E. Castle over the linearity of the arrangement of the genes in or on the chromosomes in the period 1919–1920. The very multiplicity of functions of false models found in this single debate is suggestive not only of its sophistication, but also of the richness of scientific practice in the use of false models in a way that suggests that this case is not special, but probably representative of the use of models in many other disputes.

THE BACKGROUND OF THE DEBATE

Between 1910 (when Morgan [1910b] isolated and first crossed his "white eye" mutant *Drosophila*) and 1913 (when Sturtevant published his paper describing how to map relative locations of mutants on the X chromosome using the frequencies of recombination between them), the Morgan school laid out the major mechanisms of the "linear linkage" model of the location of the genes on the chromosomes. During the next decade Morgan and his colleagues elaborated and defended this model. It won many adherents, until by the early 1920s it had become the dominant view, in spite of a number of remaining unresolved questions.

Many things were explained by this model. Primary among these was the "linkage" between traits, some of which showed a nonrandom tendency to be inherited together. Two factors or traits that were linked in this way (as first discovered by Bateson, Saunders, and Punnett in 1906) violated classical Mendelian models of inheritance. They were not always inherited together, as would be expected for pleiotropic factors that produce multiple effects, such as Mendel's example of flower and seed coat color. Nor did they assort independently, as Mendel's seven factors seemed to do. Rather, they showed a characteristic frequency of association that fell in between these extremes. This and similar cases of "partial linkage" were first presented as falsifying instances for the Mendelian theory or model of inheritance. *Here, the classical Mendelian model is providing a pattern or template of expectations against which the phenomenon of "linkage" acquires a significance, as in items 4 or 5 of the above list. Since cases of "partial linkage" represent intermediates between the two classical Mendelian types of total linkage and independent assortment, their classification in this way sets up the problem in a manner that also represents an instance of item 8.* At the time that this phenomenon was first noticed in 1906, there were no theories or models to explain it, though the Boveri-Sutton hypothesis provided a good starting point, one which Morgan and Sturtevant later developed.

With two alternative alleles at each of two loci, denoted by (A, a) for the first pair and (B, b) for the second, and starting with genotypes *AABB* and *aabb* as parents, the cases of total linkage of pleiotropy, independent assortment, and partial linkage would show the following proportions among gametes going to make up F_2 offspring:

Parents: *AABB*, *aabb*

Type of linkage	Gametic types			
	AB	*Ab*	*aB*	*ab*
Total	50%	0%	0%	50%
Independent	25%	25%	25%	25%
Partial	50 − *r*%	*r*%	*r*%	50 − *r*%

The proportion *r* (which is bounded between 0 and 25%) was found to be (1) constant for any given pair of mutations or factors, and (2) different for different pairs of factors. Most importantly, (3) *r* was independent of the starting combinations of genes in the parents. (Properties 1 and 3 were actually given by Haldane [1919] as a *definition* of linkage.) Thus, if we start with the parental types *AAbb* and *aaBB* instead of *AABB* and *aabb*, the proportions of gametic types in the F_2 generation are the exact complement of the original pattern:

Parents: *AAbb*, *aaBB*

AB	*Ab*	*aB*	*ab*
r%	50 − *r*%	50 − *r*%	*r*%

Let us denote by *R* the proportion of gametic types not found in the parents, e.g., *Ab* and *aB* for parents *AABB* and *aabb*. (Obviously, *R* = 2*r*.) Something is causing reassortment of factors in *R*% of the cases to produce new gametic combinations. Furthermore, this proportion is not a function of what genes are found together in the parents, since the complementary proportion of *AB* and *ab* gametes is found if we start with parental types *AAbb* and *aaBB*. The hypothesis of the Morgan school (Morgan 1911) was that the homologous chromosomes were winding around one another and then separating, exchanging corresponding segments.

In Figure 2.1, factor *C* is separated from the others (*A*, *B*, and *D*) through such an intertwining and separation. Morgan suggested that factors which were further apart would be separated more frequently through such a separation. This explains the different values of *r* or *R* for different pairs of factors. Factors with a constant linear location along the chromosome would always separate with the same characteristic frequency. This explains the constant proportion of new types produced by both (*AABB*, *aabb*) and (*AAbb*, *aaBB*) parental combinations. Finally, if the factors (and their alleles) kept the same relative locations along the chromosome through such interchanges, *r* or *R* should be constant for a given pair of factors and independent of what other specific alleles are found at other locations along the chromosome.

By 1913, six mutations had been localized to the X chromosome, and Sturtevant noted that the recombination frequencies between pairs of factors were approximately additive. Thus, of the frequencies of recombinant types between factors, *R*(*AB*), *R*(*BC*), and *R*(*AC*), one of them was usually equal to or slightly

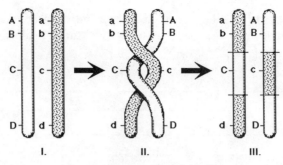

Figure 2.1. Crossing over and recombination between homologous chromosomes.

less than the sum of the other two. This suggested to him that he could use the basic intuitions of Morgan's 1911 paper to construct a map of their relative locations on the X chromosome from the recombination frequencies between all pairs of the mutations, according to successive applications of the following basic scheme. If three factors, *A, B,* and *C,* are arranged in a line, then one of them is between the other two. Which one is in the middle can be determined by seeing which of the following two conditions holds:

	Equation	Arrangement
		$A \quad B \quad C$
(2.1a)	$R(AB) + R(BC) = R(AC)$	●----●------●
		$A \quad C \quad B$
(2.1b)	$R(AB) - R(BC) = R(AC)$	●----●------●

With this scheme of mapping and Morgan's proposed mechanism for the reassortment of different factors on different chromosomes, Sturtevant could not only give the relative locations of genes on the chromosomes, but also could explain a new phenomenon, which became obvious once the factors were ordered in this way. While one or the other of the above two equations is always met approximately, sometimes the largest recombination frequency is slightly less than the sum of the other two. These deviations from strict additivity in the observed recombination frequencies between different factors are systematic. They occur only for larger recombination frequencies and increase in magnitude with the size of the recombination frequencies. *This is an instance of the fourth function of false models given above.* It involves first of all a recognition of a causal factor (relative location in the chromosome map) which causes behavior like that described in the above equations. Secondly, it involves a recognition of deviations from this behavior which, thirdly, are attributed to the action of a causal factor not taken into account in the simple model.

Suppose that crossovers occur at random with equal probability per unit distance along the length of the chromosome map. Then for relatively short distances, in which we should expect only one crossover event, as between A/a and B/b or C/c in Figure 2.1, the proportion of observed recombination events should increase linearly with the map distance. But if one is looking at a pair of factors that are sufficiently far apart that two or more crossovers have occurred between them, then the observed recombination frequency should be an *underestimate* of the map distance between the two factors. This is because a double crossover between two factors (as with A/a and D/d) brings the factors A and D back to the same side so that they are included in the same chromosome. At this early stage in the development of their model, they had misidentified the stage in meiosis when crossing over occurs, but the basic argument remains the same today. Under these circumstances, if one is not tracking any factors between the two factors in question, an even number of crossovers will be scored as no crossovers, since the factors will come out on the same side, and an odd number of crossovers will be scored as one crossover, since the factors will come out on opposite sides. Thus an increasing proportion of recombination events will be missed as the distance between the two observed factors increases.

With this bias, factors that are further separated should show increasing deviations between observed recombination frequency and actual map distance. In 1919 Haldane gave a mathematical analysis of the problem and derived from first principles what has come to be known as the "Haldane mapping function," according to which the relation between observed recombination frequency R and map distance D is given by the equation

(2.2) $$R = 0.5(1 - e^{-2D})$$

Haldane characterizes the mechanism underlying this equation as assuming an "infinitely flexible chromosome" (1919, p. 299). He also describes the case in which multiple crossing over does not occur (producing an exact proportionality between recombination frequency and map distance) as assuming a "rigid" chromosome (p. 300). He also suggests an equation for behavior between these two extremes as representing the case of a stiff or partially rigid chromosome. This kind of mechanical argument underlying the derivation of a mathematical model is extremely useful, since it suggests sources of incorrectness in the idealizing assumptions of the models, and serves to point to specific factors that should be relaxed or modified in producing better models. *As such it exemplifies item 8 from the list of functions of false models.* Haldane does not have a preferred choice for how to model the behavior of a "partially rigid" chromosome, so the best that he can do is to present two largely unmotivated models for the intermediate case, one of which is an analytically tractable but rather *ad hoc* model and the second of which is explicitly just a "curve-fitting" attempt for use in prediction.

The Haldane mapping function was either unknown to or at least was never used by the members of the Morgan school during the period of the debate with Castle over the linear arrangement of the genome in 1919 and 1920. This was possibly because their empirically determined "coincidence curves" for the various chromosomes, which gave frequency of double crossovers as a function of map distance, indicated the interaction of more complex and variegated causal factors determining multiple crossing over (see Muller 1916a, 1916b, 1916c, 1916d).

Whatever the explanation for their failure to use Haldane's model(s), they clearly had a sound qualitative understanding of the predicted behavior of their proposed mechanism from the first. Such an understanding is displayed already in Sturtevant's 1913 paper, where he pointed out that with the proposed mechanism, there should be a very small frequency of double crossovers even between factors that are quite close together. These could be detected by tracking a factor that was between the two factors in question (as C is between A and D in Fig. 2.1). But he detected no close double crossovers, and the frequency of crossovers between relatively distant factors was substantially less than expected if one crossover had already occurred in that interval. (The relatively distant factors were 34 map units apart, where one map unit is that distance for which 1% recombination is expected.) He therefore hypothesized an "interference effect" acting over a distance around a crossover within which a second crossover was prevented or rendered exceedingly unlikely. *Here is yet another application of the "template-matching" function of models of item 4, where the deviation from the expected performance of the template is used to postulate another causal factor whose action caused the deviation.*

This additional hypothesis of an "interference effect" was required for the model of the Morgan school to account for the exact linearity found in map relations for close map distances (see Fig. 2.4). This was the subject of extensive investigations by Muller in 1915 and 1916, who considered the mechanisms of crossing over, proposed a number of alternative mechanical hypotheses to explain this effect, and sought to gather data to choose among them (Muller 1916a, 1916b, 1916c, 1916d). The mechanical model for recombination was a fruitful source of hypotheses on the cause of this phenomenon. For example, if we assume that chromosomes have a certain rigidity and torsional strength, it is clear that they can be wound up only so tightly (like a double-stranded rubber band) before they break. Thus chromosome rigidity and torsional strength determine a maximum tightness of the loops before breakage would occur. This is turn determines a minimum length of chromosome between crossover events—an "interference distance" (see Haldane 1919).

The occurrence of interference also had a beneficial secondary effect, since it meant that they could use recombination frequencies between closely linked factors as a direct measure of map distance, rather than correcting them for unobserved multiple crossovers. (If recombination events occurred at random along the chromosome map, and without interference, so that they were statistically independent, then the multiplicative law for the cooccurrence of multiple events would apply, and for a distance in which there was a probability p of one crossover, there would be a probability p^2 of two crossovers, p^3 of three crossovers, p^n of n crossovers, and so on.)

To appreciate the character of the debate with Castle, it is important to realize the theoretical character of the parameter "map distance." One could not infer from a chromosome map exactly where the factors were on the chromosome. As Sturtevant noted in his 1913 paper, the chromosome might have differential strengths and probabilities of breakage along its length, leading in general to a nonlinear relation between map distance and distance along the chromosome. Nor was it possible to determine which end of the map corresponded to which end of the chromosome without the production of aberrant chromosomes having visibly added or subtracted segments at an end, something which was not done until the 1930s.

In the case of *Drosophila*, the debates continued on a theoretical plane until the (re)discovery by Painter in 1934 of the giant salivary gland chromosomes, which had banding patterns whose presence or absence and arrangement could be visibly determined. This allowed the location of genes at specific banding patterns and the ready detection of inversions, reduplications, translocations, and deletions. Some of these were hypothesized earlier solely on the basis of changed linkage relations—a remarkable (and laborious) triumph that makes the earlier work on chromosome mapping one of the most elegant examples of the interaction of theoretical and experimental work in the history of science. The subsequent localization of genes to bands, by studying genetic deletions, also permitted confirmation of Sturtevant's conjecture that the mapping from genetic map to chromosome, while order-preserving, was nonlinear because the frequency of crossing over was not constant from one unit of chromosome length to the next (see Fig. 2.2).

CASTLE'S ATTACK ON THE "LINEAR LINKAGE" MODEL

Not everyone was enamored of the mechanical models of the Morgan School. A number of geneticists (including Bateson, Punnett, Goldschmidt, and Castle) attacked it on a variety of grounds between 1913 and 1920, and Goldschmidt was still an outspoken critic in 1940. (See Carlson's [1967] excellent history for a discussion of some of their views.) At the time they had what seemed like good reasons for these attacks, reasons which led them (and Morgan too—as late as 1909 and 1910; see Allen 1979) also to attack the Boveri-Sutton hypothesis of 1903 that the genes were located on the chromosomes. The Boveri-Sutton hypothesis was a direct ancestor of the linear linkage models of the Morgan school.

In spite of the theoretical power of the Boveri-Sutton hypothesis and the later models of the Morgan school, many developmental biologists and geneticists were bothered that these models had nothing to say about gene action, but explained only correlations in the inheritance of traits. But the tradition from Aristotle down through the beginning of the twentieth century was that the hereditary elements, whatever they were and however they worked, determined both the transmission of traits and their development. It seemed unreasonable to them that any theory that dealt only with the first of these could be correct, and they sought a single theory that would explain both (see Morgan 1909; Carlson 1967; and Allen 1979). The Morgan school was embarrassed by this but chose to defer consideration of this problem, give the many successes of their research program, and transmission genetics went on apace without having much, if anything, to say about development until the rise of molecular biology and the development of models of gene action like the operon model of the early 1960s.

Two things characterized the theoretical attempts of these opponents: (1) All of their models attempted at least to make room for gene action, usually in a variety of different ways; and (2) they all seemed to remain sceptical, not only about the Morgan model, but about any conclusions that appeared to follow too closely from it. Virtually everyone accepted the data that it explained—that seemed too hard to object to—but this was combined with a mistrust of the mechanical models of the Morgan school. A frequent line taken by geneticists, one that outlived widespread opposition to the model of the Morgan school, and discoverable

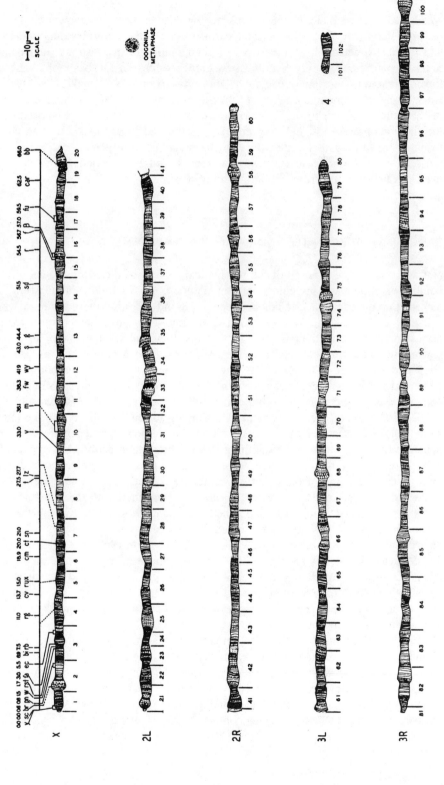

as late as Stadler (1954), was to be an "operationalist" or "instrumentalist" about the gene. Given that they could trust the data from breeding experiments, but not the models that were produced to explain the data, it seemed more reasonable to avoid commitments to the theoretical models and to accept the genes only as "operationally defined" entities that produced the breeding results—in all cases staying as close to the data as possible.

One reason why Castle's operationalist attacks (1919a, b, c) on the model of the Morgan school is so interesting is that the subsequent debate provides a clear example of the superiority of a mechanical or realist research program over an operationalist or instrumentalist one. Some of the reasons why this is so will be apparent in the following discussion, though many others will have to be left for another occasion.

In 1919 Castle published his first attack on the "linear linkage" model. He found the linear model and the associated theoretical concept of "map distance" to be too complicated and too far from the data. He was suspicious anyway of the linear model, and complained that "it is doubtful . . . whether an elaborate organic molecule ever has a simple string-like form" (1919a, p. 26). He was bothered even more by the fact that map distances for three of the four *Drosophila* chromosomes exceeded 50 units, which he saw as inconsistent with the fact that observed recombination frequencies never exceeded 50%. This apparent conflict was a consequence of his "operationalist" assumption that map distance should be made proportional to recombination frequency and represents a misunderstanding on his part of what the Morgan school was claiming. In addition, however, to construct the map from recombination frequencies one had to invoke the possibility of multiple crossing over and posit "interference effects," and Castle regarded both of these assumptions as dubious *ad hoc* additions made by the Morgan school to defend their model.

Given his "operationalist" preferences, Castle suggested that the simplest hypothesis would be to assume that the distance between the factors (whatever it signified) was a linear function of recombination frequency. He proceeded to construct a mechanical model of the arrangement of the genes in the chromosome by cutting wire links with lengths proportional to the recombination frequencies and connecting them, producing the phenomenological model of the X chromosome of *Drosophila* diagrammed in Fig. 2.3. He thereby claimed to have produced a model that fit the data and did so without making the further hypotheses of double crossing over and interference effects. In spite of his "operationalist" stance, he went on to suggest and use two alternative (and mutually inconsistent) mechanistic models to interpret the significance of his construction and to attack the linear linkage model. Like most of the other operationalist opponents of the Morgan school, Castle seems to have been a frustrated mechanist!

The force of the counterattacks of the Morgan school was to show that (1) the data Castle used did not support his model; (2) he did not use the data acceptably for a test of the linearity hypothesis; (3) his model would fare even worse if more

---

Figure 2.2. Linkage map of factors in the X chromosome and their corresponding bands in the physical (salivary gland) chromosome. (Sturtevant and Beadle 1939, Fig. 48, pp. 130–131.)

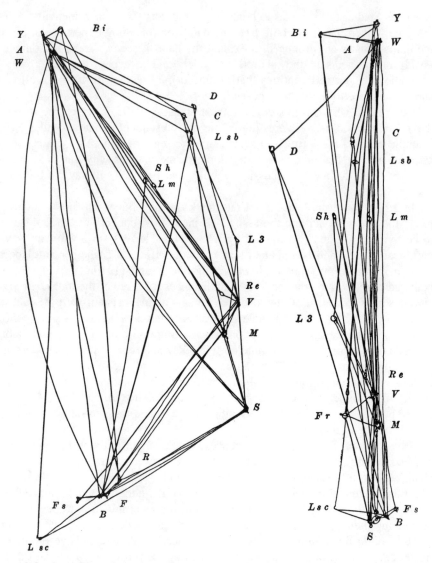

Figure 2.3. Castle's nonlinear model derived by making distances proportional to recombination frequencies, showing relative positions of genes of 20 sex-linked characters. Left-hand diagram shows side view of model, right-hand diagram shows edge view. (Castle 1919a, p. 29.)

data were included; and (4) neither of the two alternative mechanisms he proposed to account for his phenomenological model would have the desired effects.

Castle's critique and a later one were the subject of responses by Sturtevant et al. (1919) and Muller (1920). Muller's response covered virtually all of the ground of the earlier authors and went much further. His critique is elegant, covering not only the selection of the data and the reasons why it had to be treated differently for map construction than for testing the hypothesis of linear arrangement, but also an explanation of why Castle's model worked as well as it did in spite of its incorrectness, predictions that it could be expected to break down and where, a defense of the reasonableness (and inevitability) of interference and multiple crossing over, and a complex set of arguments that Castle's claims were not only inconsistent with the data but internally inconsistent as well.

MULLER'S DATA AND THE HALDANE MAPPING FUNCTION

In his 1920 paper, Muller uses data derived from his own earlier experiments (Muller 1916a, 1916b, 1916c, 1916d) to argue for the inadequacy of Castle's model. These data are of the observed frequencies of recombination between each possible pair of six of the factors on the X chromosome. In the model of the Morgan school, the recombination frequencies between nearest neighbor factors are assumed to be the map distances, and deviations from additivity for frequencies of recombination between more distant factors (see discussion of Sturtevant above) are seen as products of multiple crossovers between the observed pair of factors. The greatest distance between nearest neighbors among the factors Muller chose is 15.7 map units, and the lowest distance for which nonadditives are observed is 23.7 map units. Thus the assumption that nearest neighbor distances are real map distances, and not underestimates as would be produced by multiple crossing over, seems justified. Another problem emphasized by both Castle and Muller is that the rarity of recombination events between very closely linked factors could make very small distance estimates unreliable. Muller obviated this objection by picking six factors (out of 12 for which he had data) that were neither too closely nor too distantly spaced along the chromosome map.

In Figure 2.4 I have graphed the relation between recombination frequency and map distance to be expected in two cases: (1) the linear relation between the two supposed by Castle's model, and (2) the curvilinear relationship expected for the Haldane mapping function (HMF). Notice that with the HMF, the relationship for small distances is approximately linear but deviates from linearity with increasing map distance and asymptotes for very long distances at 50%. This suggests that factors very far apart on a long chromosome are essentially uncorrelated in their inheritance. This is a new prediction of the linkage analysis, since it suggests that two factors can be uncorrelated in their inheritance *either* by being on different chromosomes (as before) òr by being sufficiently far apart on a chromosome with a long map.

Both of these curves are to be compared with Muller's data, which suggest a relationship that remains essentially linear for much larger map distances than with the HMF (as a product of interference effects), but that also appears to approach a recombination frequency of 50% asymptotically for very great map distances as with the Haldane mapping function. This actually cannot be seen in Muller's data, since the most distant factors he lists do not have very large separations. It can be seen for the data in Haldane's graph (1919, p. 309), which interestingly also shows much greater scatter in the points—probably a product of the fact that Haldane took data from different experiments. As Muller (1920) argued (against

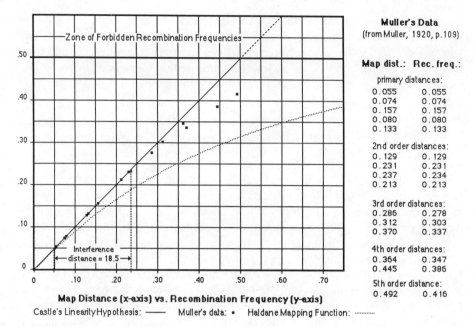

Figure 2.4. Relation between recombination frequency and map distance according to Castle's linearity assumption and the Haldane mapping function with Muller's data (1920, p. 109) and an empirical estimate of interference distance shown.

Castle), this practice may show confounding effects due to changing linkage relationships produced by different translocations, inversions, or deletions in the stocks used in the different experiments, different temperatures, and other factors that affect recombination frequency.

Notice several things here:

1. The actual data Muller gives behave qualitatively like the HMF, and might well be mistaken for it if the HMF were not graphed along with it for comparison, pointing to the systematic discrepancies. The HMF is being used here as a template, deviations from which point to the importance of interference, which is a causal factor affecting the data but not taken of in the mechanisms supposed in deriving the HMF. It is being a knowingly used as a model that is false because it leaves out a known causal factor, which produces systematic and increasing deviations from the predictions of the HMF with increasing map distance. *This is an instance of the fourth function of false models.*

2. If the data curve did not look qualitatively like that of the HMF, there would be less temptation to use it as a template and more temptation to say that it fundamentally misrepresented the mechanism involved, rather than being false by being merely incomplete. In that case the partitioning of the effects of causal factors along the lines suggested by the model could not be trusted. Given their qualitative similarity (in particular, the linearity of R and D for small distances and the asymptotic approach of R to 50% as D gets very large), this temptation is strong, however. The tendency to treat the HMF as a "baseline" model is further increased by the fact that Haldane presents two other models, one for "rigid"

chromosomes, in which R is supposed to be linear with D, at least up to 50%, and one for "semi-rigid" chromosomes, in which interference has an effect intermediate between the two extremes. Muller's data also fall in between these two extremes (though not exactly on the curve of Haldane's intermediate model) and are thus plausibly treated as due to the joint operation of factors found in the HMF model together with the operation of some kind of interference mechanism. (*See function 8 of false models in the above list.*)

3. The discrepancies between the HMF and the data can actually be used to get a rough estimate of the interference effect. Thus, if one observes where Muller's data begin to deviate noticeably from the line $R = D$ (at 23.4) and estimates where the HMF begins to deviate comparably (at about 5), one can get a rough estimate of interference distance as the difference of these two numbers—about 18 or 19 maps units. I subtract 5 from 23.4 to correct for the fact that double crossovers at close separations may be present in too small a frequency to be detected. One can thus use the "measurability lag" for the HMF to estimate that for the real data. Here an incorrect model is being used to calculate a correction factor which is then applied to real data, after which the difference between the incorrect model and the real data is used to estimate the magnitude of effect of a causal factor which is not included in the incorrect model. *This is a more complicated instance of the fifth function of false models given in the above list.* This estimate is a brute empirical estimate from the behavior of the data, not one from any assumptions about the mechanism or mode of action of interference, and as such is a procedure that is fraught with further dangers. Only with a model of the mechanism of interference does it become trustworthy.

Interestingly, the mechanisms of interference have proven since to be quite refractory to analysis in terms of general equations that both apply to a variety of cases and are derived from underlying mechanical models—probably because the mechanisms both are mathematically complex to describe and vary substantially from case to case. Thus, for example, the location of the two factors relative to the centromere is a crucial determinant of their recombination behavior, a factor that is not considered in any of these models. As a result, many more recent treatments, such as those of Kosambi (1944), Bailey (1961), and Felsenstein (1979), have constructed phenomenological mathematical models with no mechanical underpinning, but which have other advantages. These include (1) providing a relatively good fit with known data, (2) providing a schema for prediction, (3) generating important prior models (such as Haldane's) as special cases, (4) having nice formal properties, and (5) producing nice operational measures that can be applied to more complex breeding experiments. Felsenstein (1979) explicitly mentions the last four of these, and *the third is an instance of the eleventh function of false models.* With possibly unintended irony, Felsenstein summarizes the advantages of his phenomenological model over other more realistic approaches:

> There are a number of papers in the genetic literature in which specific mapping functions are derived which are produced by particular models of the recombination process. . . . While these models have the advantage of precision, they run the risk of being made irrelevant by advances in our understanding of the recombination process. In this respect the very lack of precision of the present phenomenological approach makes it practically invulnerable to disproof (1979, p. 774).

Felsenstein's more extended defense illustrates part of the ninth function of model building (predictive adequacy), but also claims that the lack of underlying mechanical detail may make it more robust in the face of new theory (see also Cartwright 1983). If this model is more immune to falsification, this "advantage" is purchased at the cost of a descent into the possible abuses of "curve fitting," which Felsenstein warns us against by explicitly noting the "phenomenological" character of his model. As I will show, however, being a phenomenological model is no necessary guarantee against falsification.

MULLER'S "TWO-DIMENSIONAL" ARGUMENTS AGAINST CASTLE

Castle's model generates a three-dimensional figure because the recombination frequencies between more distant factors are less than the sum of recombination frequencies between nearer factors, a feature earlier referred to as "nonadditivity." This produces triangular figures for each triple of factors, which, when connected together, produce complex polyhedral structures in three dimensions, as in Figure 2.3. If we look at three factors at a time, A, B, and C, and arrange them so that they are at distances proportional to their pairwise recombination frequencies, the fact that $R(AC) < R(AB) + R(BC)$ implies that A, B, and C will be at the vertices of a triangle. (This relationship is the "triangle inequality" of plane geometry.) As the deviation gets smaller, the extreme angles in the triangle get smaller, producing for $R(AB) + R(BC) = R(AC)$ the "degenerate triangle" of a straight line. Only if $R(AB) + R(BC) < R(AC)$ is it impossible to construct a triangular figure with straight edges.

If we consider another factor D, such that $R(BC) + R(CD) > R(BD)$, we get another triangle, which must be placed on top of the first, since they share side $R(BC)$. The addition of this triangle to the first generates a new distance $R(AD)$, which provides a prediction of the recombination frequency between A and D that can be compared with the data. In this way, with successive applications of this construction, Muller constructed a map of the six factors according to Castle's principles. Surprisingly, his map is two-dimensional, rather than three-dimensional as Castle required! Why was it legitimate to use a figure of lower dimensionality, which would thereby give up the use of an additional degree of freedom for fitting the data?

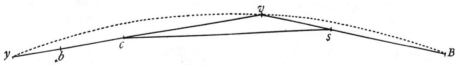

Direct representation of the strongest and second strongest linkages in Table II. (y bi, bi cl, cl v, v s, s B, and y cl, bi v, cb s, v B, are each represented by a line of length proportionate to the respective frequency of separation.) The dotted curve shows the "average angular deviation" of the line of factors, according to this system.

Figure 2.5. Muller's construction of the map of the six factors, y, bi, cl, v, s, and B, using Castle's method and the closest and next closest distances. (Muller 1920, Fig. 3, p. 113.)

FALSE MODELS   45

1. Muller argues (1920, p. 113) that his two-dimensional figure is the only possible one using the "strongest and second strongest" linkages—i.e., using the R values for nearest neighbor and next nearest neighbor factors. This is true for a reason that he does not explicitly mention: Because he has used relatively tight linkages (sufficiently close to prevent double crossovers in all cases but one), the "triangle inequality" holds for only one triple of factors—for all others there is an equality. In this special case, we get a "degenerate" figure which is only two-dimensional, because there is a "bend" at one factor and all of the other relations are linear. Two noncolinear straight lines connected at one point determine a plane. The addition of any more data that had nonadditivities would have required at least three dimensions.

2. The argument that Muller emphasizes most strongly is that the distance between the most distant factors (with four factors in between) on the model constructed according to Castle's principles is too great—it is 49.3 as opposed to his observed recombination frequency of 41.6. Muller notes that the predicted distances are also too great for the factors that are three or two factors apart. He claims that if the figure is bent so as to make the longest distance correct (his Fig. 4, p. 113), then the shorter distances become too short, and the model still fails to fit the data. Thus, as he argues earlier," The data . . . could not be represented either in a three-dimensional or in any other geometrical figure" (p. 112).

3. We can go further than Muller does, as might be suggested by Muller's references in two places to higher-dimensional figures. If there were nonlinearities or bends at two places, but they were sufficiently small that the predicted recombination frequency between the two most distant factors was still too large, then going to three dimensions (i.e., by rotating a part of the figure out of the plane) would not help Castle. This could serve only to *increase* the distance between the factors, and thus the error in his predictions, if they had already been geometrically arranged in two dimensions so as to minimize the distance and predictive errors. If the predictions had been systematically too small, then going to a three-dimensional figure would help, but if they were already too large, as they were, then nothing would work. Even if the greatest distance were too small, one would still have to fit the shorter distances and, if they are supposed to represent physical distances, one is limited to a maximum of three dimensions. In any case, this procedure would obviously be nothing more than a "curve (or polyhedron!) fitting" technique, as Muller's arguments suggest.

Here we see arguments in which simpler models (a two-dimensional model with a reduced data set) are used to draw conclusions about more complex ones (three-dimensional models with larger data sets). In particular, the argument is that given the data in question, no geometrical figure (in any number of dimensions) could consistently represent the data as a set of distances without bending wires or other such "cheating." (Note that Castle's model (in Fig. 2.3) does contain several bent wires, which represent incorrect predictions, an inconsistency noted by Muller.) *This is an example of the sixth function of false models, in which a simpler false model is used to argue for the inadequacy of a family or more complex related models.*

COUNTERFACTUAL USES OF FALSE MODELS

We can take one more step, in this case going beyond the actual dispute, to illustrate a different kind of use of false models to counter a claim of Castle that he could do without the hypothesis of interference distance. Muller argued that Castle's triangular models produced predictions for recombination frequencies that were too great at larger distances. If we transform Muller's data (counterfactually) to produce data that would be expected in the absence of interference, the fit of the data with Castle's model is far worse even than before, with much larger errors in the opposite direction. Thus Castle cannot complain about the Morgan school's use of interference effects, since this additional argument shows that he needs them even more than they do.

I pointed out earlier that interference was very useful to the Morgan school, since it allowed them to assume that recombination frequencies between factors that were fairly close together on the chromosome map were true measures of their relative map distances and not biased underestimates, as would be the case if multiple crossovers between the factors were possible. Let us drop this assumption and assume that close distances also are subject to the systematic bias produced by multiple crossovers. Then the "true" map distances (in a world with "infinitely flexible" chromosomes, and thus no interference) could be calculated from the HMF, and these new transformed distances could be used to construct a chromosome map according to Castle's model. Such a map is constructed in Figure 2.6 using Muller's data transformed according to the HMF.

In this two-dimensional figure there is a bend at each factor, since the triangle inequality is satisfied for each triple of nearest neighbor factors because of the nonlinearities introduced by the (supposed) multiple crossovers. These nonlinearities make the map bend so strongly that the predicted recombination frequen-

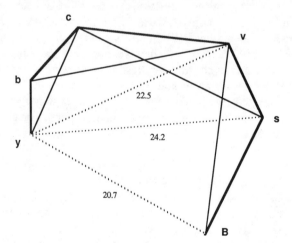

Figure 2.6. Two-dimensional map constructed according to Castle's principles using Muller's data transformed according to the Haldane mapping function to estimate the effect of assuming no interference. (Primary distances or nearest separations are shown by heavy lines, secondary distances by light lines, and predictions of higher-order distances from y by dotted lines.)

cies for more distant factors actually *decrease* as the map curves around on itself. (Compare distances ys and yB.) This data transformation demonstrates that without interference, Castle's model would have been even more strongly inconsistent with the data. Three dimensions would have helped Castle in this case, but unless the irregular helix that would be produced for a long chromosome by bending in a third dimension was very elongated, one would have had a distance function which oscillated for increasingly distant factors, one again inconsistent with data available at the time. Furthermore, a choice of other intermediate factors would have given different relative locations for the two factors at the ends. Castle could, of course, have bent successive distances in alternate directions or according to some other still more irregular scheme, but would have had to do so in a way that was consistent with all of the other shorter distances, which are not used in this demonstration. At best, the construction would have been very *ad hoc,* and the addition of any new factors would have produced new problems.

In this argument, the use of false models is very circuitous indeed. True data are transformed according to a false model (the HMF) to produce false data, which are then plugged into a false two-dimensional but simpler version of a false three-dimensional model to argue that the worsened fit of that model with the data undercuts Castle's attack on the Morgan school's use of interference by showing that he needs it even more than they do. This is an *ad hominem* argument, but one that is entirely legitimate, since it uses false premises that Castle wishes to accept. It is perhaps closer to a *reductio ad absurdum* argument, since it involves accepting a premise that they wish to reject to undercut against a premise that they wish to accept. *This example illustrates the seventh listed function of false models.* It is contrived, in the sense that they did not actually make this argument, though it is certainly one that they could have made with the data and models at hand, and it would not have been far from the spirit of either Haldane's analysis or of Muller's to do so. This case is closer to the spirit of many cases of "neutral" model building, where constructed ("random") data sets are used to test hypotheses about selection.

FALSE MODELS AS PROVIDING NEW PREDICTIVE TESTS THAT HIGHLIGHT THE IMPORTANCE OF FEATURES OF A PREFERRED MODEL

In 1917 Goldschmidt published a paper that criticized the "linear linkage" model and proposed an alternative explanation for the factor exchanges found in cases of crossing over. In his account, factors were tied to specific locations on the chromosome through biochemical forces. Corresponding allelic factors left the chromosome to accomplish their activities in the cell, an explicit attempt to accommodate gene action. They then returned to their places for the chromosomes to undergo any mitotic or meiotic activities. Each factor had a specific "force" that acted with its binding site, and different alleles had somewhat different forces. The differences between forces for alleles at the same locus were much less than the differences between any one of them and the forces of factors at other loci. When they later reassembled on the chromosome, the similarity of allelic forces with their respective binding sites resulted in occasional interchanges between them. The differences between nonallelic factors were assumed to be too great for nonallelic interchanges. The greater the differences between allelic factors the stronger would be the (apparent) "linkage" of those factors with the others on the same chromosome, since they would have less chance of going to the binding site on the homologous chromosome.

Sturtevant (1917) argued that the pattern of occurrence of multiple crossovers ruled out Goldschmidt's proposed explanation. On Goldschmidt's model, factor exchanges between alleles at one locus should be uncorrelated with factor exchanges between alleles at any other loci, since their occurrence in each case was simply a product of the similarity of alleles at the same locus, and should be independent of any events at any other loci. On the model of the Morgan school, the intertwining and linear linkage that produced factor exchange suggested that if a factor crossed over then other factors near it should also cross over. Thus, if we start with homologous chromosomes:

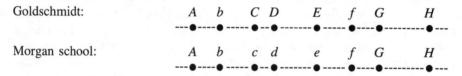

the predicted outcomes of multiple crossovers (looking only at the first chromosome) would be qualitatively as follows:

Goldschmidt:

Morgan school:

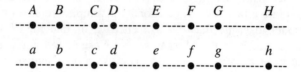

The data available to them clearly supported the "linear linkage" model.

Bridges (1917) made another crucial criticism. He pointed out that on the model of Goldschmidt, factors that had a frequency of crossing over of 1% in one generation, and whose configuration of forces was stable, should have a 99% frequency of "crossing back" to their original location in the next generation. This was at odds with the stability of crossover frequencies for a given pair of factors, and in particular with the independence of these frequencies from the parental types used in the original cross. (See the discussion of linkage, particularly property 3 noted above.)

Of these two predictions (both discussed in Carlson 1967), the first appears not to have been made before the appearance of Goldschmidt's model. (Actually, Muller in 1916 [a–d] apparently referred to the model, attributing it to "some geneticists," the Morgan school may have had knowledge of the Goldschmidt model prior to publication, since *Genetics* was a "house" journal for them at that time.) The second made use of an already known property of linkage phenomena, but in effect gave this property a greater importance. *These thus illustrate the third function of the use of false models—to generate new predictive tests of or to give new significance to features of an alternative preferred model.*

FALSE MODELS AND ADAPTIVE DESIGN ARGUMENTS

One other common use of false models would seem to be special to evolutionary biology and to other areas, such as economics and decision theory, where optimization or adaptive design arguments are common. Where a variety of alternative forms are possible, and one is looking for an explanation for why one is found rather than any of the others, it is common to build models of the various possible alternatives and to specify fitness functions of properties of their structure so that the fitnesses of the various alternatives can be estimated and compared. *When the fitnesses of the other alternatives are significantly less than that of the type found in nature, we have a* prima facie *explanation for why they are not found—at least if they are assumed to arise only as variants of the given type, or it from them.* We do not in this case have either an explanation for the occurrence of the given type (since we have shown only that it would be adaptive if it *did* arise, and not that it would do so), or an explanation for their nonexistence (if they can arise other than in selective competition with the given type). Finally, we may not have the correct explanation for their nonoccurrence if there are other factors that could independently prevent their evolution or render these variants maladaptive.

This strategy of explanation can be illustrated with a simple thought experiment which provides an explanation for why there are no species with three or more sexes. (This topic was the subject of a paper in the early 1970s by A. D. J. Robertson which I read at that time but have been unable to find since, and remember at best unclearly. Thus, some, all, or none of this argument may be due to Robertson.)

We must first distinguish two things that could be meant by more than two sexes. It might mean several sexes of which any two could have fertile offspring, but mating was biparental. This system could be called "disjunctive," since there is a variety of possible types of successful matings. Alternatively, it might mean that three (or n) different types of individuals are all required to produce fertile offspring. This system could correspondingly be called "conjunctive," since only one mating type is successful, and this requires a conjunction of n individuals.

The first type of system is uncommon, but is found occasionally. The wood-rotting fungus *Schizophyllum communis* has a variety of mating types, mates biparentally, and almost any two types will do. There are two genetic complexes, A and B, with 40 to 50 variants at each "locus." To be compatible types, two individuals must have different variants at each of the two loci (see King 1974). The second type of system does not exist in nature. There are a variety of considerations that could explain why this is so, and I will consider only one of them.

I will assume the simplest kind of interaction among members of a species, in which matings are modeled on collisions (or on linear reaction kinetics). Let an individual have probability p of encountering another reproductively potent individual during the time when it can mate successfully. This probability is a partial measure of fitness, which is presumably a product of it with several other factors. With a two-sex system, if the sexes are equally frequent there is a probability of $p/2$ that an individual can have a pairing that will produce offspring—one with the opposite sex. With a disjunctive n-sex system, where any two can play if they are of different sexes, the probability of finding a mate becomes

$[(n - 1)/n]\, p$, since an individual can mate with any other sex, and all other sexes constitute a proportion $(n - 1)/n$ of the individuals it could meet. This quantity is greater than $p/2$ for $n > 2$, which shows that systems with more sexes have an advantage. This could be one factor explaining the multiplicity of types in *Schizophyllum communi*, and raises the interesting question why it is not more common. Presumably there are other constraints in addition to those considered in this model.

It is also worth mentioning that Goldschmidt's model required a large array of distinguishable forces which all met the condition that allelic forces were more alike than nonallelic ones. This requirement raised potential problems for mutations that produced sometimes substantial phenotypic effects without changing their linkage relations, a fact noted later by Muller, who used this to argue that the mechanism of gene transmission must be independent of the mechanism of gene expression. It would also undercut the advantages for the arrangement of factors in chromosomes first hypothesized by Roux in 1883 and widely accepted since then. In Roux's account, chromosomes were seen basically as devices to increase the reliability of transmission of factors to daughter cells in mitotic division by putting many factors in one package, and thus reducing the number of packages that had to be assorted correctly to transmit each factor in mitosis. The linear chromosomal organization of factors gave a simple mechanical solution to the requirement of getting a complete set of genetic factors without having to take care of each and every factor as a special case, while the Goldschmidt model restored the need for special forces and arrangements for each factor.

A "conjunctive" multisex system represents an even more interesting kind of thought experiment, for it leads to an explanation of why such systems have *not* been found in nature. If one sex of each type must get together during their reproductive period, the probability of this occurring is $(p/n)^{n-1}$, since an individual would be expected to meet individuals of each other sex in the relevant period with a probability of (p/n) per sex, and there are $(n - 1)$ other sexes. (I assume that meetings are statistically independent. Note also that if the sexes are not equally frequent, the probability of the right conjunction occurring is even smaller, so that in this respect, this is a "best case" analysis.) $(p/n)^{n-1}$ is always less than $p/2$ for $n > 2$, and very much smaller if p is much less than 1. If $p = 1/2$, an individual of a two-sex species has a chance of $1/4$ of finding a mate, an individual of a conjunctive three-sex species has a chance of $1/36$ of finding its two mates of the right sexes, and an individual of a conjunctive four-sex species has a chance of only $1/512$ of finding its necessary three mates of all the right sexes. Note that slight extensions of these arguments could be deployed against more complex mixed "disjunctive-conjunctive" systems, such as one in which any k out of n sex types could reproduce conjunctively, since whatever the value of n, $k = 2$ is selectively superior to any larger value of k. A classic puzzle of sexuality is also preserved in this model, since an asexual species obviously does better in this respect than any of these more complex systems, including the familiar two-sex system.

This brutally simple model thus serves to explain why there are no conjunctive three-or-more-sex mating systems. Of course, there are other reasons: The classic work of Boveri (1902) on the catastrophic consequences of simultaneous fertilizations of a single egg by two sperm shows that a variety of other constraints point in the same direction. Without a great deal of reengineering of the mitotic and meiotic cycles, it is hard to imagine how a conjunctive three-or-more-sex system could ever evolve from a two-sex system. The mitotic and meiotic cycles are deeply "generatively entrenched": They are effectively impossible to modify because so many other things depend upon their current forms. This is a paradigm of the kind of developmental constraint I discussed earlier. (See also Arthur 1984; Rasmussen 1986; Wimsatt 1986; and Schank and Wimsatt, in preparation.)

This almost playful example illustrates a point which is of broad importance in evolutionary biology and equally applicable in feasibility or design studies in engineering. In the latter case, much effort is expended to model the behavior of possible systems to determine whether they should become actual. If they are found wanting, they never get beyond design or modeling stages because the proposals to build them are rejected. The "false" models that describe their behavior are never found in nature, not because they make approximations or are idealizations that abstract away from nature (though they are surely that!), but because their idealizations represent *bad designs* that are maladaptive.

Obviously false models have an important role in this kind of case. It is partially for this reason that much good work in theoretical evolutionary biology looks more than a little like science fiction. This is nothing to be ashamed of: Thought experiments have a time-honored tradition throughout the physical sciences. The main advantage that evolutionary biology has is that the greater complexity of the subject matter makes a potentially much broader range of conditions and structures the proper subject matter of biological fictions.

Part of what makes such biological fictions interesting is that it is not always possible to tell whether a model is a fiction because of the necessary idealizations required to get models of manageable complexity or because of selection for superior alternatives. A nice example of this kind of situation is provided by the recent discussions of the importance of chaotic behavior in the analysis of ecological systems. I reviewed much of this literature (Wimsatt 1980a) and argued that the absence or rarity of chaotic behavior in ecological systems did *not* show that avoidance of chaos was not a significant evolutionary factor. If, as seemed likely, there were evolutionary possible adaptive responses open to the relevant species that could act to control or avoid chaotic behavior, we would expect selection to incorporate these changes because chaotic behavior is generally maladaptive. There is not only the possibility of at least local extinction (as widely noted in the literature), but also the fact that major fluctuations in population size, sex ratio, and effective neighborhood can easily generate much smaller effective population sizes and lead to substantially reduced genetic variation (Wimsatt and Farber, unpublished analysis). This could in turn lead either to increased probability of extinction of "chaos-prone" populations or species in changing environments, or to their more rapid evolution via mechanisms suggested in Wright's (1977) "shifting balance" theory.

Furthermore, after a number of years in which the common wisdom has seemed to be that chaotic behavior was rare in nature, more recent work (e.g., Schaffer 1984) suggests that chaotic behavior may be much more common than we suspect and can be seen if we use the right tools to look for it. Also, the kinds of biases discussed in Wimsatt (1980a) would tend to hide chaotic behavior, even in cases where we have the good data over an extended period of time necessary to apply Schaffer's analysis, so it may be more common than even he suggests.

Another dimension in the use of false models which I have not discussed here is the suggestion of Richard Levins (1966, 1968) that the construction of families of idealized and otherwise false models could allow the search for and evaluation of "robust theorems"—results that were true in all of the models and thus independent of the various false assumptions made in any of the models. (This is the motivation for his comment, "Our truth is the intersection of independent lies.") I have discussed this approach at length (Wimsatt 1980a, 1981a), and Taylor (1985) provides a new study in depth of the use of this approach in ecological modeling, as well as a more sophisticated analysis of the strengths and limitations of different alternative approaches.

SUMMARY

Neutral models in biology represent "baseline models" or "null hypotheses" for testing the importance of efficacy of selection processes by trying to estimate what would happen in their absence. As such they often represent the deliberate use of false models as tools to better assess the true state of nature. False models are often used as tools to get at the truth. In this chapter I have tried to analyze and illustrate the variety of ways in which this can be done, using as examples a variety of cases from the history of genetics, drawing on the development of the "linear linkage" model of the Morgan school, and the different ways in which this model was used in countering contemporary attacks and competing theories. Philosophers of science often complain that most models used in science are literally false. Some of them then go on to argue that "realist" views of science in which arriving at a true description of nature is an important aim of science are fundamentally mistaken. I have tried to illustrate the variety of ways in which such deliberate falsehoods can be productive tools in getting to the truth, and thus to argue that these philosophers have despaired too quickly. Scientists need not, in any case, be automatically defensive when someone complains that their models are false. False models (like true ones) are often misused, but I hope that I have shown that establishing their falsity is no automatic path to this conclusion.

ACKNOWLEDGMENTS

I thank James Crow and Bruce Walsh for discussions and references on modern theories and models of interference phenomena. Nils Roll-Hansen and an anonymous referee made a number of helpful suggestions. Janice Spofford and Edward Garber gave guidance on "nonstandard" mating systems. I would also like to thank a decade of students in my biology class, "genetics in an evolutionary perspective," who have acted as sounding boards for some of these ideas and as guinea pigs for the computer labs that grew out of them. Bill Bechtel, Jim Griesemer, Ron Laymon, Bob Richardson, Sahotra Sarkar, Leigh Star, Leigh Van Valen, and Mike Wade gave useful comments at earlier stages of this analysis.

REFERENCES

Allen, G. 1979. *Thomas Hunt Morgan: The Man and His Science*. Princeton: Princeton University Press.
Arthur, W. 1982. A developmental approach to the problem of variation in evolutionary rates. *Biological Journal of the Linnean Society* 18:243–261.
Arthur, W. 1984. *Mechanisms of Morphological Evolution*. New York: Wiley.
Bailey, N.T.J. 1961. *Introduction to the Mathematical Theory of Genetic Linkage*. Oxford: Oxford University Press.
Bateson, W., E. Saunders, and R.C. Punnett. 1906. Report III: Experimental studies in the physiology of sex. *Reports to the Evolution Committee of the Royal Society*. 3:1–53.
Boveri, T. 1902. On multipolar mitosis as a means of analysis of the cell nucleus (Über mehrpolige Mitosen als Mittel zur Analyse des Zellkerns. *Verhandlungen der physikalisch-medizinischen Gesellschaft zu Würzburg* 35:67–90). Reprinted in *Foundations of Experimental Embryology*, 2nd ed., 1974, ed. B.H. Willier and J.M. Oppenheim, pp. 74–97. New York: Macmillan.
Bridges, C.B. 1917. An intrinsic difficulty for the variable force hypothesis of crossing over. *American Naturalist* 51:370–373.
Carlson, E.O. 1967. *The Gene: A Critical History*. Philadelphia: W.B. Saunders.
Cartwright, N. 1983. *How the Laws of Physics Lie*. London: Oxford University Press.
Castle, W.E. 1919a. Is the arrangement of the genes in the chromosome linear? *Proceedings of the National Academy of Sciences, USA* 5:25–32.
Castle, W.E. 1919b. The linkage system of eight sex-linked characters of *Drosophila virilis*. *Proceedings of the National Academy of Sciences, USA* 5:32–36.
Castle, W.E. 1919c. Are genes linear or non-linear in arrangement? *Proceedings of the National Academy of Sciences, USA* 5:500–506.
Felsenstein, J. 1979. A mathematically tractable family of genetic mapping functions with different amounts of interference. *Genetics* 91:769–775.
Glymour, C. 1980. *Theory and Evidence*. Princeton: Princeton University Press.
Goldschmidt, R. 1917. Crossing-over ohne Chiasmatypie? *Genetics* 2:82–95.
Haldane, J.B.S. 1919. The combination of linkage values and the calculation of distance between the loci of linked factors. *Journal of Genetics* 8:299–309.
Kauffman, S.A. 1969. Metabolic stability and epigenesis in randomly constructed genetic networks. *Journal of Theoretical Biology* 22:437–467.
Kauffman, S.A. 1985. Self-organization, selective adaptation, and its limits: A new pattern of inference in evolution and development. In *Evolution at a Crossroads: The New Biology and the New Philosophy of Science*, ed. D.P. Depew and B.H. Weber, pp. 169–207. Cambridge, Mass.: MIT Press.
King, R.C. 1974. *Bacteria, Bacteriophages and Fungi—Handbook of Genetics. Vol. 1*. New York: Plenum.
Kosambi, D.D. 1944. The estimation of map distances from recombination values. *Annals of Eugenics* 12:172–176.
Levins, R. 1966. The strategy of model-building in population biology. *American Scientist* 54:421–431.
Levins, R. 1968. *Evolution in Changing Environments*. Princeton: Princeton University Press.

Lewontin, R.C. 1978. Adaptation. *Scientific American* 239:212–230.
Morgan, T.H. 1909. What are 'factors' in Medelian explanations? *Proceedings of the American Breeder's Association* 5:365–368.
Morgan, T.H. 1910b. Sex-limited inheritance in *Drosophila*. *Science* 32:120–122.
Morgan, T.H. 1911. Chromosomes and associative inheritance. *Science* 34:636–638.
Muller, H.J. 1916a-d. The mechanism of crossing over I-IV. *American Naturalist* 50:193–221; 284–305; 350–366; 421–434.
Muller, H.J. 1920. Are the factors of heredity arranged in a line? *American Naturalist* 54:97–121.
Painter, T.S. 1934. A new method for the plotting of chromosome aberrations and the plotting of chromosome maps in *Drosophila melanogaster*. *Genetics* 19:175–188.
Rasmussen, N. 1986. A new model for developmental constraints as applied to the *Drosophila* system. *Journal for Theoretical Biology*, in press.
Roux, W. 1883. On the significance of nuclear division figures: A hypothetical discussion (Über die Bedeutung der Kerntheilungsfiguren). *The Chromosome Theory of Inheritance*, 1968, ed. B. Voeller, pp. 48–53. New York: Appleton-Century Crofts.
Schaffer, W.M. 1984. Stretching and folding in lynx fur returns: Evidence for a strange attractor in nature? *American Naturalist* 124:798–820.
Schank, J., and W.C. Wimsatt. 1986. Generative entrenchment and evolution. In *Proceedings of the 1980 Biennial Meeting of The Philosophy of Science Association, Vol. 2*, ed. P.K. Machemer and A. Fine, pp. 1–55. Lansing, Mich.: The Philosophy of Science Association.
Simon, H.A. 1981. *The Sciences of the Artificial*, 2nd ed. Cambridge, Mass.: MIT Press.
Stadler, L.J. 1954. The gene. *Science* 120:811–819.
Sturtevant, A.H. 1913. The linear arrangement of six sex-linked factors in *Drosophila*, as shown by their mode of association. *Journal of Experimental Zoology* 14:43–59.
Sturtevant, A.H., 1917. Crossing over without chiasmatype? *Genetics* 2:301–304.
Sturtevant, A.H., and G.W. Beadle. 1939. *An Introduction to Genetics*. Philadelphia: W.B. Saunders. Reprinted 1962. New York: Dover Books.
Sturtevant, A.H., C.B. Bridges, and T.H. Morgan. 1919. The spatial relation of genes. *Proceedings of the National Academy of Sciences, USA* 5:168–173.
Taylor, P.J. 1985. Construction and turnover of multi-species communtites: A critique of approaches to ecological complexity. Ph.D. dissertation, Department of Organismal and Evolutionary Biology, Harvard University.
Williams, G.C. 1966. *Adaptation and Natural Selection: A Critique of Some Current Evolutionary Thought*. Princeton: Princeton University Press.
Wimsatt, W.C. 1980a. Randomness and perceived-randomness in evolutionary biology. *Synthese* 43:287–329.
Wimsatt, W.C. 1980b. Reductionistic research strategies and their biases in the units of selection controversy. In *Scientific Discovery. Vol. 2: Case Studies*, ed. T. Nickles, pp. 213–259. Dordrecht: Reidel.
Wimsatt, W.C. 1981a. Robustness, reliability and overdetermination. In *Scientific Inquiry and the Social Sciences*, ed. M. Brewer and B. Collins, pp. 124–163. San Francisco: Jossey-Bass.
Wimsatt, W.C. 1981b. Units of selection and the structure of the multi-level genome. In *Proceedings of the 1980 Biennial Meeting of the Philosophy of Science Association, Vol. 2*, ed. P.D. Asquith and R.N. Giere, pp. 122–183. Lansing, Mich.: The Philosophy of Science Association.
Wimsatt, W.C. 1985. Forms of aggregativity. In *Human Nature and Natural Knowledge*, ed. A. Donagan, N. Perovich, and M. Wedin, pp. 259–293. Dordrecht: Reidel.

Wimsatt, W.C. 1986. Developmental constraints, generative entrenchment, and the innate-acquired distinction. In *Interdisciplinary Cooperation Among the Sciences,* ed. P.W. Bechtel, pp. 185–208. Amsterdam: North-Holland. In press.

Wright, S. 1977. *Evolution and the Genetics of Populations. Vol. 3: Experimental Results and Evolutionary Deductions.* Chicago: University of Chicago Press.

3

Self-Organization, Selective Adaptation, and Its Limits: A New Pattern of Inference in Evolution and Development

STUART A. KAUFFMAN

Current evolutionary theory is in a state of healthy flux. Our general framework for considering the relation between ontogeny and phylogeny remains dominated by the Darwinian-Mendelian marriage. Within that conceptual marriage, Mendelian segregation and mutations which are random with respect to fitness provide the mechanisms to maintain the population variance upon which selection acts. Selection is viewed as providing disruptive, directional, or stabilizing forces acting on populations, leading to phenotypic divergence, directional change, or stasis (Simpson 1943). The early dominant conceptual reliance on selection as the preeminent evolutionary explanatory principle has led to debates about the coherence of a panadaptationist program in which all features of organisms might, in some sense, be maximally adapted (Gould and Lewontin 1979). The discovery of abundant genetic variance within and between populations has led to debates about the extent to which random drift and fixation versus disruptive, directional, or stabilizing selective forces determine population genetic behavior (Lewontin 1974; Ewens 1979). However, the possibility that divergence, directional change, or stasis in the face of arbitrary mutational events might be partially due to the constraining self-organized features of organisms (Rendel 1967; Waddington 1975; Bonner 1981; Rachootin and Thomson 1981; Alberch 1982; Wake et al. 1983) has not been incorporated as a truly integral part of contemporary evolutionary theory.

The relative inattention to the contribution of the organism towards its own evolution, that is to "internal factors" in evolution, finds one of its conceptual roots in the very success of the Darwinian-Mendelian heritage, which replaced the typological paradigm of the rational morphologists, who sought underlying universal laws of form (Webster and Goodwin 1982), with an emphasis on population biology and continuity by descent. In turn this shift has led to the curious character of biological universals in our current tradition.

While not actually obligated by the Darwinian-Mendelian heritage, it is nevertheless true that we have come to regard organisms as more or less accidental accumulations of successful characters, grafted onto one another piecemeal, and once grafted, hard to change. This view resurfaces in our deeply held beliefs about biological universals, which appear to us as historical contingencies: accidental but typically useful properties which are widely shared by virtue of shared descent. Thus the code is universal by virtue of shared descent, vertebrates are tetrapods by shared descent, and the vertebrate limb is pentadactyl by shared descent. Yet, presumably a different code would have sufficed, and vertebrates might have had six limbs, all heptadactyl. Whether due to selection or random drift and fixation, all are widely shared but historically contingent properties. The underpinning for this view derives from the abundant evidence for descent with minor modifications. Massive changes are massively disruptive, monsters are typically hopeless, and only minor changes typically survive and pass the selection-drift filter.

It is easy to see that this tradition is at best weakly predictive, hence that phenomena such as widespread morphological stasis despite substantial genetic variance find no ready account in contemporary theory (Gould and Eldredge 1977; Schopf 1980; Wake et al. 1983; but see Charlesworth et al. 1982). Gradual change and accumulation of properties thereafter fixed is easily understood. Beyond this, and appeal to design principles and maximal adaptation, we have few conceptual tools to predict the features of organisms. One way to underline our current ignorance is to ask, if evolution were to recur from the Precambrian when early eukaryotic cells had already been formed, what organisms one or two billion years later might be like. And, if the experiment were repeated myriads of times, what properties of organisms would arise repeatedly, what properties would be rare, which properties were "easy" for evolution to happen upon, or which were "hard"? A central failure of our current thinking about evolution is that it has not led us to pose such questions, although their answers might in fact yield deep insight into the expected character of organisms.

GENERIC PROPERTIES OF SELF-ORGANIZING SYSTEMS: A SOURCE OF AHISTORICAL UNIVERSALS

Imagine, then, that evolution had recurred myriads of times, allowing a catalogue of common and rare properties. How would we try to think about the common properties? Presumably they are common by virtue of two quite different mechanisms. The common properties might reflect recurrent selection of the same useful features. Conversely, the common features might reflect properties of organisms so easily found in evolution as to be essentially unavoidable. Restated, perhaps there are properties of organisms that recur, not by virtue of selection, but by virtue of being inherent properties of the building materials. Such properties would constitute *ahistorical* universals. It is this theme I wish to explore, for I want to suggest that many aspects of contemporary organisms may reflect inherent, "generic" properties of self-organizing systems, and the *limited* capacity of selection to cause deviation from those inherent properties. Those generic properties, then, will function as something like normal forms. The core of the pattern of inference I want to consider is this: Understand the generic structural, organizational, and dynamical properties of the systems used in generating organisms; understand the extent to which selection can change those inherent properties, and then build a picture of the expected properties of organisms. This new typological pattern of inference might be useless but for these considerations: In a variety of ways, some discussed below, very powerful self-organizing properties in organisms, constituting developmental constraints, can be expected and can begin to be sketched. The limitations in the capacity of selection to modify such properties are open to investigation. Experimental implications are already available and should increase. I would note that this pattern of inference is almost completely absent in our current theories.

TWO PRELIMINARY CASES

Evolution has not recurred *ab initio,* but an alternative way to ask whether there may be generic ahistorical features of organisms consists in considering whether there are widely shared properties among distant phyla that seem highly unlikely to reflect either selection or shared descent. I briefly mention two.

Rhythmic phenomena are found in organisms at a wide range of levels. Included among these are circadian rhythms in animals and plants, neural periodicities, firefly flash rhythms, yeast glycolytic oscillations, and cardiac rhythms (reviewed in Winfree 1980). A common experimental technique to investigate all these rhythms is to apply a perturbation such as a temperature shock or light flash, then study the alteration in the phase of the rhythmic phenomenon. Application of the same perturbation at all phases results in a phase-resetting curve mapping a new phase as a function of the old phase. In his recent book, Winfree (1980) has argued that such phase-resetting phenomena have essentially universal properties that recur over and over in myriad circumstances. It is conceivable that the common properties reappear on such a wide range of levels and of phyla due to recurrent selection for similar adapted phase-resetting behavior. Winfree's own argument seems more plausible. The nearly universal properties are elegantly understood as generic topological properties of continuous oscillatory systems. Build a biochemical oscillator; it is very likely to exhibit a restricted family of phase-resetting behavior, not because of selection, but because those features are generic to such oscillations.

In all higher metazoan and metaphytic phyla, each cell type during ontogeny differentiates along branching developmental pathways directly into rather few other "neighboring" cell types. It is logically conceivable, and might be highly advantageous, for a single cell type to proliferate, then differentiate directly into all the very many other cell types found in the adult, after which those cell types would rearrange themselves into the adult morphology. Indeed, dissociated sponges are capable of this rearrangement dance (Humphreys 1963; Moscana 1963), yet normal sponge ontogeny follows typical branching developmental pathways (Willmer 1960). It might be the case that this feature of virtually all metazoan and metaphytic phyla reflects recurrent selection for a common property of high adpative advantage. I have described in detail elsewhere (Kauffman 1969, 1974, 1983) and shall suggest below a contrary thesis: Due to the self-organizing dynamical properties of complex genomic regulatory systems, any cell type can generically have but *few* accessible neighboring cell types. This is no trivial constraint. It implies that higher organisms must rely upon branching pathways of development, with a restricted number of branches at each branch point. If this view has merit, ontogenies for the past billion years have been largely constrained to utilize this pattern, not because it is adpative, but because it is essentially unavoidable.

I have chosen these two preliminary examples because the properties which appear to be generic are not trivial, but are already important organizing features of higher plants and animals. Further, one feels impelled here to admit that a selective explanation is suspect, and that the properties in question might possibly reflect self-organizing features. Yet selection surely operates, and the fundamental problem I want to discuss is how to begin to think about the relation between generic self-organizing features of organisms, the action of selection, which may attempt to modify those properties, and the *balance* struck by selection and the restoring forces of random mutation, which tend to drive the selected system back towards those properties which are generic to it; hence the *limits* of selection. An example will help make the argument more clear.

THE "WIRING DIAGRAM" OF THE GENOMIC REGULATORY SYSTEM

In a sense, the genomic system in a cell is loosely analogous to a computer. A mammalian cell has enough DNA to encode on the order of 1,000,000 average-sized proteins (Bishop 1974). Current estimates in echinoderms (Hough et al. 1975; Kleene and Humphries 1977) and vertebrates (Bantle and Hahn 1976; Hastie and Bishop 1976; Chikaraishi et al. 1978) suggest that, in fact, up to about 50,000 to 100,000 distinct transcripts are actually found in the heterogeneous nuclear RNA, and a subset of these are processed to the cytoplasm as mature message. In eukaryotes, as in prokaryotes, the genomic system includes structural genes, which code for proteins, and both *cis*-acting and *trans*-acting regulatory genes, which control the expression of structural and other regulatory genes. The modes of action and varieties of regulatory genes are still only partially known. However, it is already clear on genetic and direct molecular evidence in yeast (Sherman and Helms 1978; Errede et al. 1980; Struhl 1981), mice (Paigen 1979; Kurtz 1981), *Drosophila* (Lewis 1978; Dickenson 1980a, 1980b; Corces et al. 1981), and maize (McClintock 1957; Peterson 1981), that *cis*-acting sites regulate the transcriptional activity of structural genes in some local domain on the same chromosome. *Trans*-acting genes typically lie at distant positions on the same or other chromosomes, and presumably exert their influence on a given structural gene indirectly, via diffusible products that interact with *cis*-acting sites near the structural gene (Britten and Davidson 1971; Abraham and Doane 1978; Davidson and Britten 1979; Paigen 1979; Dickenson 1980a; Green 1980; Kurtz 1981). Regulation of gene expression occurs at the transcriptional level, in the processing of heterogeneous nuclear RNA to the cytoplasm as mature messenger RNA, in the translation to protein, and in posttranslational modifications of the protein that modulate its activity (reviewed by Brown 1981). The loose analogy of this complex system to a computer lies in the fact that the 100,000 or so different genes and their products form a network whose components regulate one another's activities. Thus, a cell is a dynamical system, whose complete accounting would require at least specification of all the particular regulatory interactions; that is, a specification of the "wiring diagram" showing which components affect which components, and the local "rule" for each component describing the behavior of that component based on the behaviors of the components directly affecting it. Familiar examples of such genetic "circuitry" include the *lac* operon in *Escherichia coli* (Zubay and Chambers 1971) and the complex regulatory system in bacteriophage lambda (Thomas 1979).

For the purpose of discussion, I wish to idealize the very complex system presumed to exist in eukaryotic cells by which one gene influences the expression of another gene. I shall idealize the genomic regulatory system by imagining that the genome consists of structural genes whose transcription is regulated by nearby *cis*-acting genes; that each *cis*-acting gene regulates all transcribable genes in some local domain which is demarcated in some way; that *trans*-acting genes lie in the domain of *cis*-acting genes, which regulate the expression of the adjacent *trans*-acting genes, while the *trans*-acting genes themselves produce products that are targeted to act on specific "matching" *cis*-acting sites anywhere in the chromosome set.

This crude, but current picture of genetic regulation in eukaryotes is captured in Figure 3.1a. Here I have assumed that the genome has four kinds of genetic elements, *cis*-acting (Cx), *trans*-acting (Tx), structural (Sx), and empty $(-)$ domain-demarcating elements. For concreteness, I have assumed a haploid organism with four chromosomes, having these kinds of genes dispersed in the chromosome set. Any *cis*-acting site is assumed to act in polar fashion on all *trans*-acting and structural genes in a domain extending to its right to the first blank locus. Each indexed *trans*-acting gene, Tx, is assumed to regulate *all* copies of the corresponding *cis*-acting gene, Cx, wherever they may exist in the chromosome set. Structural genes, Sx, are assumed to play no regulatory roles, although this assumption is not critical to the discussion. In Figure 3.1a I have arrayed 16 sets of triads of *cis*-acting, *trans*-acting, and structural genes separated by blanks on

CHROMOSOME 1   C1 T1 S1—C2 T2 S2—C3 T3 S3—C4 T4 S4—

CHROMOSOME 2   C5 T5 S5—C6 T6 S6—C7 T7 S7—C8 T8 S8—

CHROMOSOME 3   C9 T9 S9—C10 T10 S10—C11 T11 S11—C12 T12 S12—

CHROMOSOME 4   C13 T13 S13—C14 T14 S14—C15 T15 S15—C16 T16 S16—

Figure 3.1a. Hypothetical set of 4 haploid chromosomes with 16 kinds of *cis*-acting (C1, C2, ...), *trans*-acting (T1, T2, ...), structural (S1, S2, ...), and "blank" genes arranged in sets of 4: Cx, Tx, Sx, $-$. See text.

EVOLUTION AND DEVELOPMENT

CHROMOSOME 1   C1 T2 S1—C2 T3 S2—C3 T4 S3—C4 T5 S4—

CHROMOSOME 2   C5 T6 S5—C6 T7 S6—C7 T8 S7—C8 T9 S8—

CHROMOSOME 3   C9 T10 S9—C10 T11 S10—C11 T12 S11—C12 T13 S12—

CHROMOSOME 4   C13 T14 S13—C14 T15 T14—C15 T16 S15—C16 T1 S16—

Figure 3.1b. Similar to Figure 3.1a, except the triads are Cx, T(x + 1), S(x), −. See text.

the four chromosomes. A graphical representation of the control interactions among these hypothetical genes is shown in Figure 3.2a, in which an arrow is directed from each labeled gene to each gene that it affects. Thus, in Figure 3.2a, $C1$ sends an arrow to $T1$ and an arrow to $S1$, while $T1$ sends an arrow to $C1$. A similar simple architecture occurs for each of the 16 triads of genes, creating 16 separate genetic feedback loops. By contrast, in Figure 3.1b, each triad carries the indices $[Cx, T(x + 1), Sx]$, while the sixteenth is $(C16, T1, S16)$. This permutation yields a control architecture containing one long feedback loop, Figure 3.2b.

The first point of Figures 3.1 and 3.2 is that each spatial arrangement of *cis*, *trans*, and structural genes into domains along the chromosomes can be put into one-to-one correspondence with a specific "wiring diagram" showing which genes affect which genes. While this hypothetical example is obviously idealized, it is also true of the genomic system in contemporary cells. A spatial arrangement into domains of *cis*-acting, *trans*-acting and structural genes corresponds to some actual wiring diagram of regulatory interactions, for example, bacteriophage lambda (Thomas 1979). Inclusion of the complete list of interacting products of each gene as components of a eukaryotic system would correspond to drawing a more complex wiring diagram.

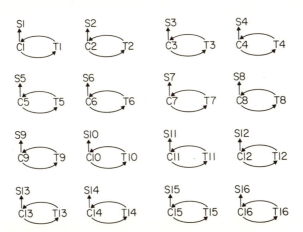

Figure 3.2a. Representation of regulatory interactions according to rules in text, among genes in chromosome set of Figure 3.1a.

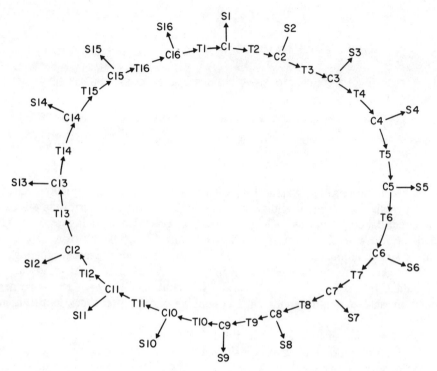

Figure 3.2b. Regulatory interactions among chromosome set in Figure 3.1b.

The second obvious point of Figures 3.1 and 3.2 is that alteration of the spatial arrangements of *cis*, *trans*, and structural genes on the chromosomes corresponds to alterations in the "wiring diagram."

THE SCRAMBLING GENOME: EXPLORING AN EVOLUTIONARY ENSEMBLE WITH GENERIC PROPERTIES

It is now widely, but not universally, thought that chromosomal mutations are an important factor in evolution (Bush et al. 1977; Wilson et al. 1977; Bush 1980; Dover et al. 1982; Flavell 1982). These mutations include not only tandem duplications which generate new copies of old genes but growing evidence which suggests that fairly rapid dispersion of some genetic elements occurs through chromosomal mutations including transposition, translocation, inversions, and conversions. Dispersal of loci by processes such as these provides the potential to move *cis*- or *trans*-acting genes to new positions and thereby create novel regulatory connections, having implications for oncogenesis (Cairns 1981) and also opening novel evolutionary possibilities.

Appreciation of the potential evolutionary import of chromosomal mutations is widespread (Roeder and Fink 1980; Shapiro 1981; Gillespie et al. 1982; Hunkapiller et al. 1982; but see Smith 1982). The new issue I want to address is this: Suppose we are able to characterize with some precision the frequencies with which different DNA regions actually duplicate, or are dispersed randomly or nonrandomly in the chromosome set. Some currently unknown subset of these alters the genomic wiring diagram. Suppose for the moment we ignore any effect of selection. Then chromosomal mutations persistently scramble the genomic wiring diagram in some way consistent with what we shall eventually discover about the actual probabilities of local duplication and dispersion. In the absence of selection, other than that which may influence the chromosomal mutation rules themselves, this scrambling process will explore the *ensemble* of all possible wiring diagrams consistent with the constraints in the scrambling process. Thus, in the absence of further selection, we would expect the persistently scrambled genomic system eventually to display a wiring diagram whose features were *typical* of the ensemble. Therefore, characterizing such typical or "generic" properties is fundamental, for they are just the properties to be expected in the absence of selection beyond the rules of the scrambling process itself. The typical properties of the "well-scrambled" genome constitute the proper null hypothesis for our further analysis. I next show that the well-scrambled genome can be expected to exhibit very highly self-organized properties.

We are not yet in a position to characterize the ensemble of genomic systems actually explored in evolution. But a simple beginning can be made with the idealized models in Figures 3.1 and 3.2. To begin study of the effects of duplication and dispersion of loci on such simple networks, I ignored questions of recombination, inversion, deletion, translocation, and point mutations completely, and modeled dispersion by using transpositions alone. For the haploid chromosome set I used a simple program that decided at random whether a duplication or transposition occurred at each iteration and over how long a linear range of loci, between which loci duplication occurred, and into which position transposition occurred.

Even with these simplifications, the kinetics of this system is complex and not further discussed here, since the major purpose of this simple model is to examine the regulatory architecture after many instances of duplication and transposition have occurred. I show the results for the two distinct initial networks in Figures 3.3a and 3.3b, for conditions in which transposition occurs much more frequently than duplication (the ratio of the frequency of transposition to that of duplication is 0.90 : 0.10), and 2,000 iterations have occurred. The effect of transpositions is to randomize the regulatory connections in the system, thereby exploring the ensemble available under the constraints on duplication and dispersion. Consequently, while the placement of individual genes differs, the overall architecture of the two resulting networks (Figs. 3.3a and 3.3b) looks far more similar after adequate transpositions have occurred than at the outset (Figs. 3.2a and 3.2b). This similarity, readily apparent to the eye, reflects the fact that both networks in Figures 3.3a and 3.3b exhibit connectivity properties that are typical in the ensemble explored by the model chromosomal mutations. These generic connectivity properties can be stated more precisely.

The oversimplified model in Figures 3.1–3.3 is still too complicated for this initial discussion. The effects of duplication and transposition are to create new copies of old genes, and by their dispersal to new regulatory domains, generate both more and novel regulatory couplings among the loci. A minimal initial approach to the study of the ensemble properties of such systems is to study the features of networks in which N genes are coupled completely at random by M regulatory connections. This kind of structure is termed a random directed graph (Berge 1962), in which nodes (dots) represent genes, arrows represent regulatory interactions.

To analyze such random networks I employed a computer program that generated at random M ordered pairs chosen among N "genes," assigned an arrow running from the first to the second member of each pair, and then analyzed the fundamental connectivity features. The resulting wiring diagrams are as follows: (1) The number of genes directly or indirectly influenced by each single gene, by following regulatory arrows tail to head, termed the regulated *descendents* from each gene. A well-known example of such direct and indirect influence is the sequential cascade of over 150 alterations of puffing patterns in *Drosophila* salivary gland chromosomes following exposure to the molting hormone ecdysone (Ashburner 1970). (2) The *radius* from each gene, defined as the minimum number of steps for influence to propagate from that gene to its entire battery of descendents. The mean radius and the variance of the radius give information on how "extended" and "hierarchically" constructed the network is. Thus, long hierarchical chains of command from a few chief genes would have high mean and variance in the radius, while a network in which each gene influenced all other genes in one step would have low mean radius and low variance. (3) The fraction of genes lying on feedback loops among the total N genes. (4) The length of the smallest feedback loop for any gene that lies on a feedback loop.

GENERIC CONNECTIVITY PROPERTIES OF REGULATORY SYSTEMS CHANGE AS A FUNCTION OF THE NUMBER OF GENES AND REGULATORY CONNECTIONS

The expected network connectivity features exhibit strong self-organizing properties analogous to phase transitions in physics (Erdos and Renyi 1959, 1960), as the number of regulatory (arrow) connections, M, among N genes increases. If M is small relative to N, the "scrambled" genomic system typically would be composed of many small genetic circuits, each unconnected to the remainder. As the number of regulatory connections, M, increases past the number of genes, N, large connected genetic circuits typically form. This "crystallization" of large circuits as M increases is analogous to a phase transition. An example, with $N = 20$ genes, and an increasing number of arrows, $M = 5, 10, 20, 30, 40$, is shown in Figure 3.4. Thus, if $M = 40$, $N = 20$, typically direct and indirect regulatory pathways lead from each gene to most genes, many genes lie on feedback loops, etc. Results are shown in Figures 3.5a–d for larger networks with 200 genes and regulatory connections, M, ranging from 0 up to 720. As M increases past N, large descendent structures arise in which some genes directly or indirectly influence a large number of genes. The mean descendent curve is sigmoidal (Fig. 3.5a). When $M = 3N$, each gene typically is directly or indirectly connected to most other genes. The mean radius of the genetic wiring diagram is a nonmonotonic function of the ratio of M to N (Fig. 3.5b). As M increases from 0, each

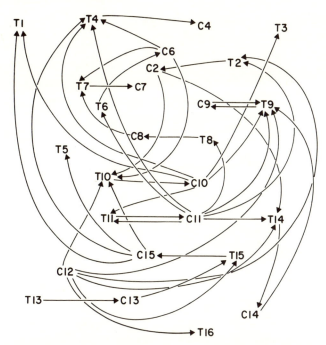

Figure 3.3a. Regulatory interactions from chromosome set in Figure 3.1a after 2,000 transpositions and duplications have occurred, in the ratio .90:.10, each event including 1 to 5 adjacent loci. Structural genes and fully disconnected regulatory genes are not shown.

gene can reach its few descendents in a few steps (Fig. 3.4). As long descendent circuits crystallize in the still sparse network, when M is only slightly greater than N, the radius becomes long. As the number of connections, M, continues to increase, new short pathways connecting a gene with its descendents are formed

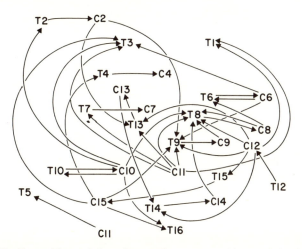

Figure 3.3b. Similar to Figure 3.3a, after random transposition and duplication in chromosome set from Figure 3.1b.

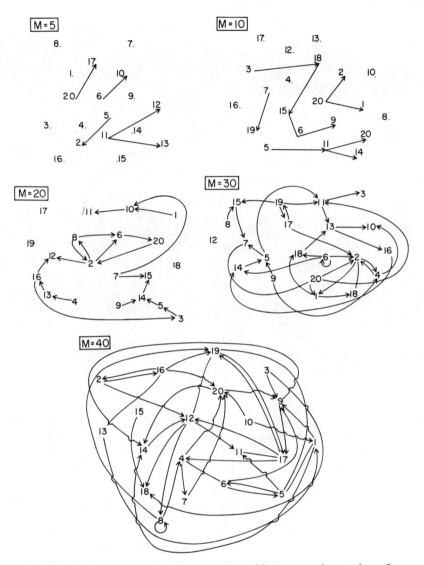

Figure 3.4. Random regulatory circuits among $N = 20$ genes as the number of connections, M, increases from 5 to 40.

and the radius gradually decreases (Fig. 3.4). The fraction of genes lying on feedback loops (Fig. 3.5c), like the number of descendents, is sigmoidal as M increases past N, since long descendent structures are likely to form closed genetic feedback loops as well. Finally (Fig. 3.5d), since the minimal lengths of feedback loops are bound above by the radius, these are also a nonmonotonic function of the $M : N$ ratio, reaching a maximum about where the two sigmoidal curves are steepest.

What shall be said of this idealized model? True genetic regulatory systems presumably do not generate novel regulatory genes and novel regulatory connec-

EVOLUTION AND DEVELOPMENT 67

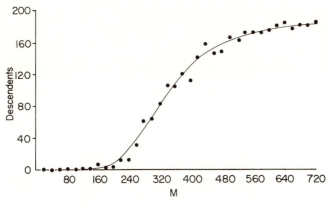

Figure 3.5a. Descendent distribution, showing the average number of genes each gene directly or indirectly influences as a function of the number of genes (200) and regulatory interactions, M.

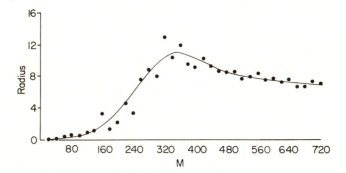

Figure 3.5b. Radius distribution, showing the mean number of steps for influence to propagate to all descendents of a gene, as a function of M.

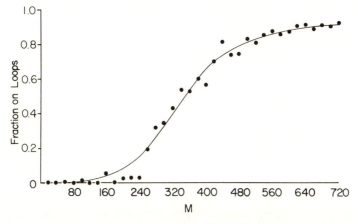

Figure 3.5c. Average number of genes lying on feedback loops, as a function of M.

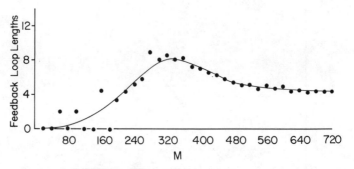

3.5d. Average length of shortest feedback loops genes lie on as a function of M.

tions in as simple a way as this primitive model. The true mechanisms shall define some more realistic and complex ensemble actually being explored in eukaryotic evolution. However, this primitive model already points out a number of fundamental facts. First, the statistical properties of such ensembles are highly robust, just as the statistical properties of fair coin-flipping ensembles are robust. Just as the expected fraction of "heads" is 50% with fair coins, so the expected connectivity properties of scrambled genetic regulatory systems are robust, and given the ensemble, characterizable. Second, the generic connectivity properties of the ensemble are not featureless, but highly structured. In the current primitive example, given 10,000 genes and 30,000 regulatory connections among them, very strong statements can be made about the expected connectivity properties of the regulatory network. To restate the now obvious point, if we can specify how chromosomal mutations actually generate new genes and novel connections, then we can characterize the ensemble actually being explored by evolution. The highly structured generic properties of the ensemble are a proper null hypothesis characterizing in detail the statistical features we would expect to find in eukaryotic genomic systems in the absence of selection. If we do not find those features, then either we have generated the wrong ensemble because we do not understand how the genome is actually scrambling, or other forces are causing the genomic wiring diagrams to deviate from generic. Chief among potential forces is selection, which I discuss next.

GENERIC PROPERTIES, MUTATION, SELECTION, LIMITS OF ADAPTATION

Underlying the approach I propose are the interrelated concepts of the statistically expected properties of the ensemble being explored in evolution, the effects of mutation and selection, and the limits of adaptation attainable in a population. These can now be preliminarily summarized. I begin by borrowing the well-known Maxwell's demon from physics. Consider two isolated and initially isothermal gas-containing boxes joined by a window where the demon allows faster molecules into the right box. As that box becomes warmer, its increased pressure counters the demon's efforts. If the demon is of finite and modest efficiency, in due course the back pressure equals his efficiency and the system comes to a steady state displaced modestly from thermodynamic equilibrium. The deviation from equilibrium reflects the *limited* capacity of the demon to select the faster molecules into the right box. Further, if the steady state is close to thermodynamic equilibrium, many statistical properties of the selected system will be explicable from the statistical features of the unselected equilibrium ensemble properties.

The demon analogy carries directly into the evolutionary problem. The generic properties of the genetic ensemble, with respect to the "wiring diagram" described above, are fundamental, for those properties are exactly the (equilibrium) features to be expected in the absence of selection. The generic properties are simultaneously available in the ensemble to be exploited by selection; yet also, since mutations drive partly selected systems back toward generic properties, those generic properties are a constraint to be overcome by selection. In turn, mutations drive systems toward generic properties, but also supply the population variance on which selection can act. Selection is analogous to Maxwell's demon, able to enrich the representation of the more desired genomic systems in the population, until the back pressure toward generic properties due to mutation balances the effect of selection. That balance is the *limit achievable* by selection. If the balance lies close to the unselected mean, then the underlying generic properties of the ensemble will in large measure account for the properties we see. Then within the ensemble actually explored in evolution its highly structured generic statistical properties will function as normal forms or ahistorical universals. A concrete example is discussed next.

POPULATION SELECTION FOR A UNIQUE ARBITRARY WIRING DIAGRAM AND ITS LIMITS

A central issue for this ensemble pattern of inference is whether selection is sufficiently powerful to achieve and maintain any arbitrary unique wiring diagram in the ensemble, throughout the population, for its properties might deviate extensively from the generic properties of the ensemble. My aim in this section is to show that, for fixed selection rules, as the complexity of the wiring diagram to be maintained in the population by selection increases, a limit is reached beyond which a unique, precisely chosen system cannot be maintained and the maximally selected network falls ever closer to the unselected mean properties of the ensemble. Hence, for sufficiently complex genetic systems, the ensemble pattern of inference should be useful. Beyond patterns of inference, however, this is a fundamental biological question about the limits of the capacity of selection to attain high precision and complexity in genetic regulatory systems in the face of mutational disruption of the selected system.

To start investigating this question, I have begun with the simplest class of population selection models, based on the Fisher-Wright paradigm (Ewens 1979). Imagine a population of initially identical haploid organisms with N regulatory genes connected by a total of T regulatory arrows. Suppose that some arbitrary specific wiring diagram assigning the T arrows among the N genes is "perfect," and that at each generation, mutations alter the wiring diagram by probabilistically reassigning the head or tail of arrows to different genes. Assign a relative fitness to each organism which depends upon its deviation from the perfect wiring diagram, such that the probability of leaving an identical offspring in the next discrete generation is proportional to relative fitness.

Perhaps the simplest form of this general model is

$$(3.1) \qquad W_x = (1 - b)\left(\frac{G}{T}\right)^\alpha + b$$

W_x is the relative fitness of the xth organism in the population, G_x is the number of good connections among its components, T is the total number of regulatory connections, b is a basal fitness, and α is an exponent that corresponds to the three broad ways fitness might correlate with the fraction of good connections. For $\alpha = 1$, fitness is linearly proportional to the number of good connections. For $\alpha > 1$, fitness drops off sharply as the number of good connections, G, falls below T. This corresponds intuitively to the case where virtually all connections must be correct for proper functioning, and any defect yields sharp loss of function. For $0 < \alpha < 1$ in Equation 3.1, small deviations from "perfect" ($G = T$) would yield only slight loss of fitness, which would fall off increasingly rapidly for larger deviations of G below T. This corresponds to a broad region of high fitness near "perfect" with wide toleration of deviations in wiring diagrams. These three classes of fitness seem fundamental; more complex patterns would be built up from these three.

I have carried out an initial computer investigation of this model for small T and small population size. At each generation, mutations at rate μ per arrow "end" randomly reassigned the "head" or "tail" of regulatory arrows to different genes, hence altering the wiring diagrams in a population of 100 haploid organisms; the fitness of each was calculated; organisms were sampled with replacement and left progeny in the next generation with a probability equal to their fitnesses. Sampling continued until 100 offspring were chosen, thus constituting the next generation, and then the process was iterated. The initial population of organisms was identical, and either "perfect" or constructed randomly, hence "generic." I sought, in particular, the steady states of this system, reflecting stationary population distributions of the number of correct connections per organism due to the balance of mutation and selection forces.

A general implication of this simplest model is that for sufficiently simple wiring diagrams, selection is powerful enough to maintain the population near any arbitrary "perfect" wiring diagram, but with increasing complexity of the wiring diagrams, the steady-state balance struck between selection and mutational back pressure shifts toward generic. In Figure 3.6a, I show simulation results for the evolution of the mean number of good connections per organism in a population of 100 organisms, each with 20 genes connected by 20 arrows. Over 1,000 generations, a population that initiated at the prefigured "perfect" wiring diagram gradually decreased the mean number of good connections, $\bar{G}$, from 20 to about 10. By contrast, a population of initially identical organisms constructed at random with respect to the "perfect" wiring diagram, hence with an average of about one good connection, gradually increased the mean number of good connections above this generic value, to about 10. Thus, under these conditions, populations

EVOLUTION AND DEVELOPMENT

exhibit a single globally stable stationary distribution with $G = 10$, reflecting the balance between mutation and selection.

As the number of connections, T, to be specified increases, the stable stationary distribution shifts smoothly from "perfect" closer to generic. Figure 3.6b shows results from simulations in populations as T ranges through values from 2 to 24, but all other parameters remain fixed. The effect of increasing the mutation

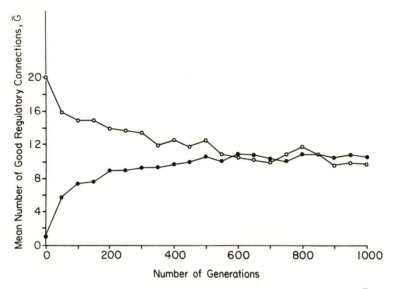

Figure 3.6a. Population selection for 1,000 generations after initiation at $\bar{G} = T$, and $\bar{G} = 1$. $\bar{G}$, mean number of good connections per organism. Simulations are based on Equation 3.1 in text, with $\alpha = 1$, $b = 0$, $P = .95$, and the mutation rate $\mu^* = .005$ per "head" or "tail" of an arrow causing a change in connection.

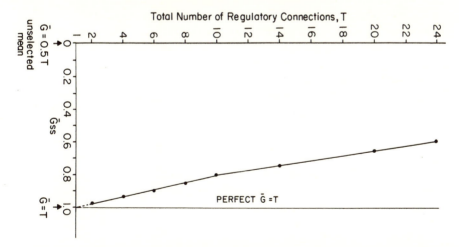

Figure 3.6b. Proportional shift of selected stationary state, $\bar{G}_{ss}$, from $\bar{G} = T$, labeled 1.0, toward unselected mean $\bar{G} = .5T$, labeled 0, as T increases. Values of $\mu^* = .005$, $P = .5$, $\alpha = 1$, $b = 0$, were used in equation 3.1.

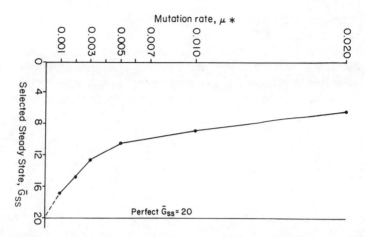

Figure 3.6c. Shift of selected steady state, $\bar{G}_{ss}$, toward unselected mean, $\bar{G} = 1$ as μ^* increases. $T = 20$, $P = .95$, $\alpha = 1$, $b = 0$ were used for simulations based on Equation 3.1.

rate, μ, was also investigated. With all other parameters fixed and low mutation rates, selection can maintain any arbitrary "perfect" wiring diagram in all members of the population, while increasing the mutation rate smoothly shifts the globally stable stationary state closer to "generic" (Fig. 3.6c).

MULTIPLE STATIONARY STATES BETWEEN "GENERIC" AND "PERFECT"

Simulation results on this simple model suggest that populations have a *single* globally stable stationary distribution under two general parameter conditions, $\alpha = 1$ or $b = 0$. Here increasing the complexity of the network to be maintained causes a smooth shift of the selected stationary distribution toward the unselected generic properties of the ensemble. In contrast, for an appropriate choice of parameters outside these conditions, the selection system can display two stationary distributions, one close to "perfect," one close to "generic," with fluctuation-driven transitions between them. Figure 3.7 shows the temporal behavior of a population (with $\alpha = 10$ and $b = .50$) initiated in the "perfect" state, which remains in that steady state for 1,000 generations, then undergoes a rapid transition to the steady state near the "generic" state, $G = 1$, where it then remains.

It is particularly interesting that this model exhibits what might be called a complexity catastrophe. As the complexity of the network to be maintained, that is, T, increases, selection can maintain the population in the nearly perfect stationary state until it decreases in stability and entirely disappears, causing the population to jump suddenly to the remaining single globally stable steady state close to generic. Thus a small increase in the complexity to be maintained, T, can cause a catastrophic change in the properties of the maximally adapted system attainable (Fig. 3.8).

EVOLUTION AND DEVELOPMENT

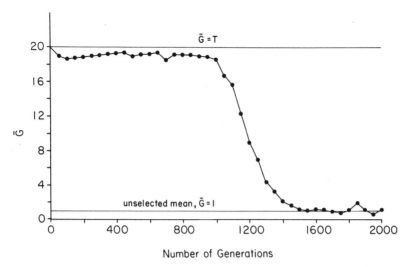

Figure 3.7. Selection system exhibits two stationary states near "perfect," $\bar{G} = T$, and unselected, $\bar{G} = 1$, with fluctuation-driven transition. $T = 20$, $P = .95$, $\alpha = 10$, $b = .5$, $\mu = \mu^* = .01$ were used in simulations based on Equation 3.1.

ANALYTIC APPROACHES

Some limited analytic insight into this population selection system has been achieved. Because I have restricted attention to a haploid model, the entire variance in fitness in the population is heritable, and Fisher's fundamental theorem applies in simplified form (Ewens 1979). The rate of change of relative fitness in the pop-

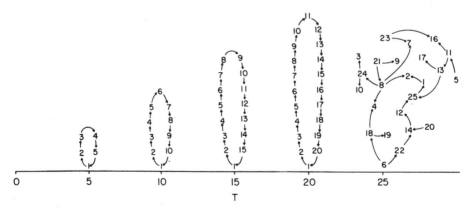

Figure 3.8. Maximally adapted wiring diagram achievable in a population by selection ($\alpha = 10.0$, $b = .5$, $\mu = 2\mu^* = .01$, Eq. 3.1) for a single closed regulatory loop. T, number of regulatory connections. By $T = 25$ the single loop cannot be maintained and the maximally adapted wiring diagram achievable falls toward the ensemble generic properties. In this simulation the number of genes, N, $= T$. Since $N = T$, and the probability of a bad connection, P, $= N - 1/N$, P changes in these simulations, but the destabilization of the nearly perfect state also occurs as T increases while P remains constant.

ulation equals the variance of fitness in the population divided by the mean population fitness:

$$\frac{d\bar{W}}{dt} = \frac{\sigma^2 W}{\bar{W}} \tag{3.2}$$

Let a total of T regulatory arrows connect the N regulatory genes. If the probability that any randomly assigned connection arrow among the N genes is a "good" connection is $(1 - P)$, and P is the probability a connection is "bad," then in a randomly connected wiring diagram, a mean $T(1 - P)$ are "good" and TP are "bad." This average reflects the generic properties of this ensemble. Mutations (at rate μ^*/arrow end/organism/generation, or μ per entire arrow $\cong 2\mu^*$) provide a restoring force (Rf) reducing the number of good connections toward this mean. The mutational restoring force is 0 at the mean and linearly proportional to the deviation in the number of good connections above the mean:

$$Rf = \mu(TP + G - T) \tag{3.3}$$

The effect of selection is to increase the fitness according to Fisher's theorem above; thus the selective force, Se, depends upon the ratio of the variance of the fitness in the population to the mean fitness. I have been unable to obtain an analytic expression for the variance as a function of T, α, b, P, μ, and population size. However, simulation results show that for fixed values of these parameters, the variance in $\bar{G}$, $\sigma^2 G$, is approximately independent of the mean number of good connections, $\bar{G}$, itself. This approximation is used next to characterize, in terms of $\bar{G}$ and $\sigma^2 G$, the balance struck between mutation and selection for "good" wiring diagrams.

Consider the simplest fitness law, $W_x = (G_x/T)^\alpha$. One first can show that

$$\sigma^2 W \simeq \left(\frac{dW}{dG}\right)^2 \cdot \sigma^2 G \tag{3.4}$$

where the derivative is evaluated at $\bar{G}$. The approximation is good if $\sigma G/T$ is reasonably smaller than 1, as observed.

The temporal derivative of $\bar{G}$ due to selection, restated in terms of $\bar{G}$ and $\sigma^2 G$, is approximated by

$$\frac{d\bar{G}}{dt} \cong \frac{dG}{dW}\frac{d\bar{W}}{dt} \simeq \left[\frac{dG}{dW}\left(\frac{dW}{dG}\right)^2 \cdot \sigma^2 G\right] \bigg/ \left(\frac{\bar{G}}{T}\right)^\alpha \tag{3.5}$$

where dG/dW and dW/dG are both evaluated at $G = \bar{G}$.

For the simplest law, $W = (G/T)^\alpha$, after substitution, the temporal derivative to increase $\bar{G}$ due to selection (i.e., effective selection, Se) is

$$\frac{d\bar{G}}{dt} = \frac{\alpha \sigma^2 G}{\bar{G}} \tag{3.6}$$

The full differential equation for the rate of change of the population mean number of good connections per organism, reflecting selection and mutation, becomes

EVOLUTION AND DEVELOPMENT

$$(3.7) \quad \frac{d\bar{G}}{dt} = \frac{\alpha\sigma^2 G}{\bar{G}} - \mu(TP + \bar{G} - T)$$

For the more complex fitness law, $W_x = (1 - b)(G/T)^\alpha + b$, where $b > 0$ reflects a "basal" fitness, the effective selection force Se increasing $\bar{G}$ per unit time becomes

$$(3.8) \quad \frac{d\bar{G}}{dt} = \frac{\alpha(1 - b)\bar{G}^{\alpha-1}\sigma^2 G}{(1 - b)\bar{G}^\alpha + bT^\alpha}$$

The full differential equation becomes

$$(3.9) \quad \frac{d\bar{G}}{dt} = \frac{\alpha(1 - b)\bar{G}^{\alpha-1}\sigma^2 G}{(1 - b)\bar{G}^\alpha + bT^\alpha} - \mu(TP + \bar{G} - T)$$

For $\alpha = 1$, fitness is linearly proportional to the fraction of good connections, and equation 3.9 simplifies to

$$(3.10) \quad \frac{d\bar{G}}{dt} = \frac{(1 - b)\sigma^2 G}{(1 - b)\bar{G} + bT} - \mu(TP + \bar{G} - T)$$

The qualitative behavior of the selection system under conditions of no basal fitness, $b = 0$, Equation 3.7, and in the presence of basal fitness, $b > 0$, but for fitness linearly proportional to the fraction of good connections, $\alpha = 1$, Equation 3.10, straightforwardly parallels the demon analogy. As shown in Figure 3.9a, the mutational back pressure toward the generic properties of the ensemble, Rf, is a monotonic increasing function of the mean number of good connections per organism, $\bar{G}$. In contrast, the selective force, Se, is a monotonic decreasing function of $\bar{G}$. For parameter values where the selective force is greater than the restoring force for all values of $\bar{G}$ less than T, the two curves do not intersect in

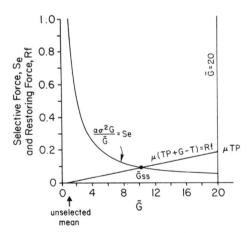

Figure 3.9a. Restoring force, Rf, tending to decrease $\bar{G}$ toward unselected mean. Selective force, Se, tending to increase $\bar{G}$, $\bar{G}_{ss}$, selected steady state. ($T = 20$, $P = .95$, $\alpha = 1$, $b = 0$, $\mu = 2\mu^* = .01$ were used in Eq. 3.1. μTP is Rf at $\bar{G} = T$.)

the interval $\bar{G} = 0$, $\bar{G} = T$, and selection is sufficiently powerful to achieve and maintain any unique, arbitrary "perfect" wiring diagram in almost all members of the populations. For parameter values where the Rf and Se curves intersect in the interval $\bar{G} = 0$, $\bar{G} = T$, the intersection corresponds to a single globally stable stationary population distribution of fitness about $\bar{G}$, and reflects the limited fitness achievable by selection due to the balancing back pressure of mutation.

These qualitative features imply that there are critical values of the parameters T, α, b, P, and μ, thus a surface in that parameter space, on one side of which selection can attain any arbitrary wiring diagram, while on the other side, the balance between selection and mutation is less than $\bar{G} = T$. Since a single steady state exists, as parameters change beyond the critical surface, the steady state will shift smoothly away from $\bar{G} = T$.

The position of the globally stable steady state depends, in particular, on the complexity of the system, T, and the mutation rate, μ. The true dependence of $\sigma^2 G$ on T and μ is unknown, but simulation suggests that $\sigma^2 T$ increases linearly with T, but less than linearly and at approximately the square root of μ. Substituting the approximation $\sigma^2 G = KT\mu^{1/2}$ into Equation 3.7 and solving for the steady state, $\bar{G}_{ss}$, yields

$$(3.11) \qquad \bar{G}_{ss} = \frac{T(1-P)}{2} + \frac{1}{2}\sqrt{[T(1-P)]^2 + \frac{4\alpha KT}{\mu^{1/2}}}$$

As either T or μ increases, the position of $\bar{G}_{ss}$ in the interval between $\bar{G} = 0$ and $\bar{G} = T$ shifts away from "perfect," $\bar{G} = T$, toward the unselected generic properties of the ensemble. These shifts for T and μ, shown in Figures 3.6b and 3.6c for the simulation results, are fit fairly well by Equation 3.11.

MULTIPLE STEADY STATES AND A COMPLEXITY CATASTROPHE

Under conditions where fitness drops off rapidly as G falls below T, that is $\alpha > 1$, and where basal fitness $b > 0$, finding analytic steady state solutions to the full differential equation (Eq. 3.9) becomes impractical. However, graphical solutions are straightforward. The following qualitative features are important in the current discussion. As shown in Figure 3.9b, for sufficiently large values of α, e.g., 10, and b, e.g., .50, the effective selective curve, Se, does not *decrease* as G increases, but *increases* at more than a linear rate and hence can cross the restoring force curve Rf at two points. The point closer to "generic," G_1, is a stable stationary state. The point closer to "perfect," G_2, is unstable and repels the population in either direction. The "perfect" state is a reflecting boundary, and therefore if the population is above the unstable steady state G_2, it becomes trapped at a nearly perfect state between the unstable steady state and perfect. Thus, the population shows two stable stationary states, with fluctuation-driven transitions between them. A time record of such a transition is shown in Figure 3.7.

The system has two stationary states. Thus, tuning the parameters increasing

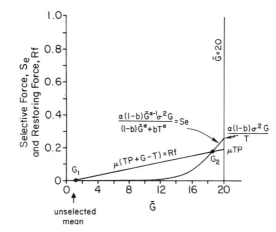

Figure 3.9b. Selective force, Se, increases with $\bar{G}$ for $\alpha = 10$, $b = .5$, $T = 20$, $P = .95$, $\mu = 2\mu^* = .01$. G_1, stable steady state; G_2, unstable steady state. μTP is Rf at $\bar{G} = T$. $[\alpha(1 - b)\sigma^2 G]/T$ is Se at $\bar{G} = T$.

mutation rate, μ, or the total number of connections to be specified, T, or decreasing the cooperativity, α, may lead to bifurcations in which the nearly perfect stationary state vanishes, catastrophically changing the maximally attainable population fitness. Explicitly, the existence of two stationary states requires that, for $\bar{G}$ close to the perfect wiring diagram, $\bar{G} = T$, the selective force be greater than the restoring force. Substituting $\bar{G} = T$ into Equation 3.9 and using the approximation $\sigma^2 G = KT\mu^{1/2}$ yields

$$(3.12) \quad \frac{d\bar{G}}{dt} = \frac{\alpha(1 - b) \cdot KT\mu^{1/2}}{T} - \mu TP$$

Existence of the nearly perfect stationary state requires $Se > Rf$, or, simplifying:

$$(3.13) \quad \frac{\alpha(1 - b)K}{P} \geq \mu^{1/2}T$$

As T increases, the selective force is roughly independent of T, while the restoring force increases proportionally to T. Therefore, as T increases, selection can at first maintain the population in the nearly perfect stationary state and gradually increase the complexity of an initial highly adapted small network, but the nearly perfect stationary state inevitably becomes progressively unstable to fluctuations then disappears. Thus, for $T = 25$ rather than 20 or less, the system in Figure 3.7 is no longer able to maintain a nearly perfect stationary state, and the population collapses to near generic (Fig. 3.8).

The maximum complexity of the wiring diagram which can be maintained by selection depends not only on the complexity, T, but on the remaining parameters. Therefore, an increase in T can be balanced by compensatory decrease in the mutation rate, μ, or increase in the selection exponent, α. However, this compensation may be both expensive and bounded. For example, Equation 3.13 suggests that μ must decrease inversely proportionally to T^2 to compensate increasing T, which is expensive, and μ will eventually reach some minimum attainable mutation rate, e.g., 5×10^{-7}. Equation 3.13 suggests for the selection system in Figure 3.7, that a maximum of perhaps 2,000 genetic regulatory connections could be maintained with precision.

Similarly, Equation 3.13 suggests an increase in T could be compensated by an increase in the selective exponent α. It is not clear if a maximum attainable value of α can be found. However, Equation 3.13 suggests that if increase in α had no effect on the variance in G, then α must increase at least linearly with T to maintain the population near "perfect." In fact, simulation results suggest that as α increases, $\sigma^2 G$ decreases, thereby reducing the effective selection force. Thus, to compensate, α may have to increase faster than T. Indeed, if as α becomes large $\sigma^2 G$ decreases linearly with increasing α, further increase in α could not compensate for increasing the complexity of the wiring diagram, which would then necessarily shift toward generic.

This study is a small first step in beginning to understand the relation between the self-organizing properties of genetic networks and selection. Further analysis of this first model of selection for a regulatory wiring diagram, and then more realistic versions, including unequal weighting of different regulatory connections, reintroduction of chromosome linkage with observed frequencies of duplication and random or nonrandom dispersion of loci, and extension to diploid models, is necessary to characterize more clearly the expected ensemble properties and the limits achievable by selection.

While still only partially explored, the qualitative behaviors of this first model underscore the fundamental biological question raised by the ensemble pattern of inference I am discussing. For other parameters fixed, and for sufficiently simple wiring diagram complexity, T, the model suggests that selection will typically be powerful enough to maintain any specific wiring diagram in almost all members of the population. Because the properties of that specifically selected system may deviate grossly from the statistically expected generic properties of the ensemble, inferences from these properties in many respects will be unwarranted. However, for other parameters fixed, as the complexity of the wiring diagram to be maintained increases, selection becomes too weak with respect to mutational forces to maintain a unique wiring diagram in the ensemble in most members of the population, and the population shifts either smoothly or abruptly to a stationary state that approaches the generic properties of the ensemble. Further, the capacity to compensate for increasing complexity by altering other parameters is probably bounded. Thus the biological issue of central importance is this: Is selection in fact sufficiently powerful that uniquely chosen genetic regulatory wiring diagrams can be maintained with minor variations in contemporary eukaryotic cells? If so, then eukaryotic regulatory systems in principle may be not only highly complex, but also precisely constructed, and thereby finely adapted to microhabitats. If not, and if the size and complexity of the regulatory network substantially exceeds the capacity of selection, then contemporary eukaryotic systems reflect some compromise between selection and the statistically robust properties of the regulatory ensemble that evolution is exploring. The implications of the latter possibility are indeed substantial. For example, considerable dispersion in wiring diagram details between close sibling species is expected, should be predictable, hence testable, and would reflect noise, not selection. More basic, as already stressed, if the balance lies close enough to the highly structured generic properties of the ensemble, then very many statistical features of eukaryotic genetic regulatory systems should be deducible from the statistical features of the ensemble and should, in fact, be observed in widely divergent groups. Thus those features should function as ahistorical universals widely present, not due to selection, but despite it. But equally important, if selection is unable to maintain both high complexity and high precision, then contemporary regulatory systems must perform adequately despite the considerable structural noise implied by being continuously "scrambled" while existing near the generic properties of the ensemble. If selection is not powerful enough to maintain any arbitrary member of the ensemble in most organisms of the population, then we may succeed in understanding major aspects of the organization of genetic regulatory systems on a statistical basis, without having to dissect the network in final detail. However, we shall have to understand the principles under which the sloppy construction maintainable by selection satisfies (Stearns 1982). Some candidate principles are discussed next.

EVOLUTIONARY COORDINATION OF PATTERNS OF GENE EXPRESSION

Even an adequate theory of the limited selective achievement of adaptive wiring diagrams would be only one step toward the more general problem of attempting to understand the evolution of adapted coordinate gene expression underlying ontogeny. The pattern of inference I wish to explore parallels the previous discussion. I shall ignore the evolution of novel structural genes and consider the still unknown regulatory system. Assuming the current fundamental hypothesis that *cis*- and *trans*-acting loci play regulatory roles, some subset of chromosomal mutations will persistently scramble the genomic wiring diagram. Presumably, some chromosomal and point mutations also modify the regulatory rules governing the behaviors of regulated genes (e.g., Dickenson 1980a, 1980b). These alterations will change the coordinated patterns of gene expression, that is, the dynamical behavior mediated by the genetic regulatory system. As in the case of the regulatory network's wiring diagram, my thesis is that we can hope to describe with increasing accuracy the ensemble of regulatory systems actually explored in eukaryotic evolution, characterize the ensemble generic dynamical behaviors coordinating the different patterns of gene expression underlying cell differentiation in ontogeny, and begin to analyze the extent to which population selection can accumulate particular adapted combinations of gene expression, and the limits achievable by selection. Again, if the limits lie close enough to the generic properties of the ensemble, then many aspects of the regulation of coordinated gene expression in higher eukaryotes may reflect, not selection, but membership in a common ensemble.

SELF-ORGANIZED DYNAMICAL BEHAVIORS OF BOOLEAN GENOMIC REGULATORY SYSTEMS

For over a decade (Kauffman 1969, 1971, 1974), I have sought to study the generic properties of genomic regulatory systems by idealizing the dynamical behavior of a single gene as a binary variable, either fully active or fully inactive. While literally false, the idealization is highly useful for two different reasons. First, it allows a description of very complex regulatory systems with, for example, 10,000 coupled components, whose dynamical behavior can be simply investigated. Second, the idealization allows precise characterization of distinct ensembles of regulatory systems which constitute initial models of the genuine ensemble currently being explored in eukaryotic evolution.

TWO LOCAL PROPERTIES OF GENOMIC SYSTEMS: LOW CONNECTIVITY AND CANALIZING FUNCTIONS

The approach I have taken is to identify two local features of genetic regulatory systems. These features are the mean connectivity of the genomic wiring diagram and the kinds of local control rules governing the activity of any gene. By connectivity, I refer to the number of genes whose actions or macromolecular products directly regulate the expression of a given gene. In the case of bacterial operons and viral genes, typically any gene is directly influenced by few (one to five) other genes (Kauffman 1971, 1974). An assessment of the local control rule governing behavior of genes in bacteria and viruses reveals an abundance of a particular class of rules I call *canalizing* rules (Kauffman 1971, 1974).

Canalizing rules are exemplified by the lactose operon in *E. coli* (Zubay and Chambers 1971). If one idealizes the operator locus as a binary variable, either free (0), or bound (1), then the behavior of the operator locus with respect to repressor molecules, and the inducer allolactose, is given by a Boolean rule. In the absence of either repressor or allolactose, the operator is free. In the presence of the repressor, but not allolactose, the operator is bound. In the presence of allolactose but not repressor, the operator is free. In the presence of both repressor and allolactose, the inducer binds the repressor and changes its configuration by pulling it off the operator, which therefore is free. This binary logic, or Boolean rule, captures the essence of the regulated behavior of the *lac* operator. This Boolean control rule has the property that if sufficient allolactose is present, the operator is free, regardless of the presence or absence of repressor molecules. I call a Boolean rule canalizing if, like the operator rule, it has the property that *one* value of some *single* regulating variable suffices to guarantee that the regulated locus assumes *one* of its two values.

The regulatory rules governing expression of eukaryotic genes are not yet known. However, a review of the known genes in bacteria and viruses reveals that the majority are governed by such canalizing Boolean functions (Kauffman 1974). It seems highly likely that the same constraint shall apply to eukaryotic systems as well. One reason to suppose so is that canalizing functions are simpler biochemically to realize than noncanalizing functions. I should stress that, while I shall

EVOLUTION AND DEVELOPMENT

consider an idealized genomic system with a unitary model of *cis* and *trans* control of transcription, the logic and methods of analysis carry over intact to a more complex and realistic system in which regulation of gene expression occurs at many levels: processing, transport, translation, posttranslational modifications, and even models where programmed genomic rearrangements occur during ontogeny. Binary models are the logical skeleton of any such regulatory system.

BIOLOGICALLY PLAUSIBLE SELF-ORGANIZED REGULATORY BEHAVIORS OF THE CANALIZING ENSEMBLE

A central feature of canalizing regulatory systems is that, even if sloppily constructed and persistently scrambled by mutational noise, such systems exhibit highly ordered regulatory behaviors.

In order to establish the generic dynamical properties of members of the canalizing ensemble, I employed simulations, sampling model genomic regulatory systems at random from this ensemble. Any such genomic system has N binary regulatory genes, each receiving regulatory inputs from a few genes, and each assigned one of the possible Boolean switching rules governing its activity. Thus, a given gene might be activated at the next moment if both of its inputs are currently active, or if any is inactive, etc.

Figures 3.10a–3.10c summarize the fundamental features of Boolean genomic models. Three genes are shown. Each possible combination of the activities of the three genes constitutes one *state* of the model genome, hence there are $2^3 = 8$ states comprising the *state space* of this system. When released from any initial state, the system follows a sequence of states through its state space, and eventually reenters a previously encountered state. Thereafter the system cycles repeatedly through this recurrent pattern of gene expression, termed a *state cycle* (Fig. 3.10c). If the net is released from a different initial state, it may follow a sequence terminating on this first state cycle, or on a distinct state cycle. Figure

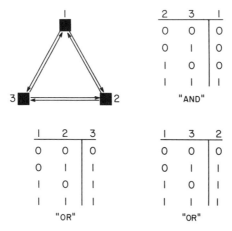

Figure 3.10a. Three genes, each regulated by the other two, according to the "OR" and "AND" Boolean function.

	T			T+1	
1	2	3	1	2	3
0	0	0	0	0	0
0	0	1	0	1	0
0	1	0	0	0	1
0	1	1	1	1	1
1	0	0	0	1	1
1	0	1	0	1	1
1	1	0	0	1	1
1	1	1	1	1	1

Figure 3.10b. All $2^3 = 8$ states of binary activity of the three genes at time T, and the state at time $T + 1$ into which each transforms.

3.10c shows that the net in Figure 3.10a possesses three alternative state cycles. The set of states flowing into, or on state cycle, constitutes its *basin of attraction*, and the cycle itself is the *attractor*. After transients die away, the alternative attractors constitute the repertoire of alternative asymptotic recurrent patterns of gene expression available to the model genomic system. Below I shall interpret each state cycle as a distinct "cell type" in the repertoire of the genomic regulatory system.

I shall discuss five critical highly ordered dynamical features of these model regulatory systems.

1. The length of state cycles. State cycles are the fundamental recurrent patterns of gene expression in these models. In principle, the number of states on the recurrent sequence might be 1, an equilibrial steady state of gene expression in the whole genome, or 2^N. State cycle lengths in the canalizing ensemble typically increase approximately as the square root of the number of genes, N. Thus, a model genomic system with 10,000 regulatory genes and $2^{10,000} = 10^{3,000}$ states settles down and cycles among a mere 100 states. Generically, then, canalizing Boolean models localize their dynamical behavior to extremely small (e.g., $10^{-2,998}$) subvolumes of their available state space.

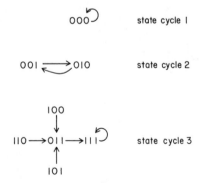

Figure 3.10c. The sequence of state transitions from Figure 3.10b showing three state cycles. See text.

I now wish to interpret such a small, asymptotic recurrent pattern of gene expression as a distinct "cell type" in the dynamic repertoire of the genome. Given this identification, the fundamental result is that, even in the absence of selection, members of the biologically plausible canalizing ensemble of Boolean regulatory systems also spontaneously exhibit sufficiently coordinated recurrent patterns of gene expression. Those patterns might constitute an adequate basis for the tightly confined coordinated gene expression required in cell types, for a single cell type stably expresses only a small number of the possible patterns of gene expression.

2. The number of alternative state cycles in the genomic repertoire of the canalizing ensemble increases as a square root of the number of regulatory genes, N, coupled in the regulatory system. Thus a genomic model with 10,000 genes would typically have on the order of *100* alternative state cycles, or recurrent asymptotic patterns of coordinated gene expression in its repertoire. If each such asymptotic attractor is interpreted as a distinct cell type, then a generic feature of this ensemble is that the number of cell types in organisms should rise as roughly the square root of the number of regulatory genes. Another way of saying this is that, as the complexity of the genome grows larger, proportionally more new regulatory genes must be added to create each additional cell type. In fact, the number of cell types among multicellular eukaryotes does increase roughly as the square root of the DNA content (Kauffman 1969, 1971). These estimates are crude because, for example, DNA content probably does not vary linearly with the number of regulatory genes, but they suffice to make several major points. First, there is, in fact, some relation between the number of cell types in organisms across widely divergent phyla and genomic complexity. This suggests the possibility that this relation does not reflect selection itself, but some other properties covarying with genomic complexity. It is possible that this correlation reflects generic features of cellular regulatory systems. Canalizing ensembles are a plausible basis for the distribution observed.

3. Stability of model cell types to perturbations is very high in members of the canalizing ensemble. Typically, any cell type is stable to about 90% of all possible perturbations reversing the activity of any single component. Thus, even in the absence of further selection, the dynamical requirements to exhibit highly homeostatic coordinated patterns of gene expression are met.

4. Branching pathways of differentiation. If a state cycle attractor is accepted as a primitive model of a cell type, then differentiation consists in transitions between attractors, triggered by appropriate signals during ontogeny. A generic feature of members of the canalizing ensemble is that any cell type can be triggered, by altering the value of any single gene, to differentiate into only a *few neighboring* cell types. Cell types are considered "neighboring" precisely because differentiation from one to another requires transiently altering the state of a single gene, e.g., by a hormonal signal. Obviously, if the states of all genes could be arbitrarily reset by a signal, any cell type could differentiate directly to any other cell type. Thus, in genetic sytems where hormonal or other signals act directly on one or a few genes, successive differentiation from any initial single cell type must follow *branching* pathways to immediate neighbors first, then their neighbors, and so on. This feature, as noted above, appears to be universal in all higher plants and animals, suggesting that the constraint for the past billion years or more

to exhibit branching pathways in development may not reflect selection, but the incapacity of selection to avoid a generic feature of the underlying genomic regulatory system.

5. A fundamental feature of members of the canalizing ensemble is that approximately 60–70% of the genes settle into fixed "on" or fixed "off" states. The dynamical consequence is simply pictured. This 60–70% constitutes a large subnet of the entire genomic system whose elements are fixed active or inactive. This large "forcing structure" subnet (Kauffman 1974) blocks the propagation of varying inactive and active "signals" from genetic loci that are not parts of the large fixed forcing structure. Those remaining genes form smaller interconnected feedback clusters, each of which is functionally isolated from influencing the other clusters by the large forcing structure (Kauffman 1974; Fogelman-Soulie et al. 1982). Each functionally isolated subsystem has a small number of alternative asymptotic dynamical patterns of behavior into which it settles: alternative steady states, or oscillatory patterns of gene activity. These feedback loop clusters therefore constitute a locus of alternative developmental decisions each capable of functioning as a master developmental switch system. Since the clusters are functionally isolated, the alternative gene expression patterns of the entire genetic system are constrained to a core consisting of housekeeping genes active in all cell types, and fixed inactive genes, plus the possible combinations of the J alternative developmental choices of the first cluster, the K choices of the second cluster, etc., and the possibly overlapping descendent batteries of genes driven by these functionally isolated master switch systems (see also Britten and Davidson 1969; Davidson and Britten 1979).

Thus, if contemporary organisms are constrained to the canalizing ensemble and exhibit its generic properties, then a number of widespread features are predicted by membership in a common ensemble. These properties include (1) underlying combinatorial logic in developmental pathways, already hinted at in *Drosophila* (Kauffman 1973; Garcia-Bellido 1975; Kauffman et al. 1978); (2) a large, ubiquitous core of genes active in all cell types as observed (Kleene and Humphries 1977; Chikaraishi et al. 1978; Ernst et al. 1979); (3) typically modest differences in patterns of total gene expression in different cell types (1–15% without further selection), again as observed (Levy et al. 1975; Axel et al. 1976; Kleene and Humphries 1977; Chikaraishi et al. 1978); and (4) cascades of alterations in gene activities signaled by reversing the activity of any single gene, which are limited to a functionally isolated cluster and its descendents, typically comprising 0–15% of the total genes. This behavior is observed, for example, in ecdysone-induced puffing activity in salivary glands of *Drosophila* (Ashburner 1970). Further generic properties concern (1) the statistically typical distribution of effects of regulatory mutations on changes in coordinated patterns of gene expression; (2) mutational generation of new cell types in evolution and their expected similarities to old cell types; (3) the existence of developmentally unused, novel, atavistic, and perhaps neoplastic cell types in the *normal* genomic repertoire; (4) the existence of "switch genes" which control transitions between normal, new, and even atavistic developmental pathways; and (5) highly nonrandom, oriented changes in developmental flow down different pathways induced by "random" mutations, suggesting that internal factors may in fact have orthogenetic consequences (Kauffman 1983).

In general, without further selection, members of the canalizing ensemble exhibit highly self-organized biologically plausible patterns of gene regulation (Kauffman 1969, 1974; Aleksander 1973; Sherlock 1979a, 1979b; Fogelman-Soulie et al. 1982). Alternative noncanalizing ensembles described elsewhere (Kauffman 1969, 1984; Walker and Gelfand 1979) do not exhibit these properties.

CONCLUSION

The canalizing ensemble of Boolean genomic models, and the random directed graph models of the genomic regulatory wiring diagram, are primitive first efforts to characterize the genuine regulatory ensembles being explored in eukaryotic evolution. My purpose in discussing them has not been to claim their final adequacy, but to characterize a new pattern of evolutionary inference using concrete examples. Although obviously preliminary, the results described suffice to demonstrate, even in the absence of further selection, that an entire *ensemble* of genomic regulatory systems exhibits remarkably self-organized coordination in patterns of gene expression. Achieving those aspects of coordination requires no selection other than membership in the appropriate ensemble. Its properties are not featureless but highly rich, stable, and statistically typical, and they would powerfully constrain and influence the style and patterns of further selective modification. It is fully reasonable to expect our characterization of genomic regulatory systems to increase in precision, and to expect that the more accurate corresponding ensemble shall also exhibit rich and stable generic properties. We have, as yet, no theory at all to assess the capacity of selection to achieve specific coordinated patterns of gene expression, its limits as the genomic system grows larger, and the constraints imposed on the style of selective modification by the underlying generic properties. We must begin to develop such theories. For, however selection acts to achieve particular choices of which genes shall be coactive in a given cell type, which housekeeping genes must be active in all cell types, which genes send or receive developmental inductive stimuli, and which cell types differentiate into which cell types, if we find that the limits of selection lie close to the statistically typical properties of the ensemble of regulatory systems actually explored in eukaryotic genomic evolution, then its highly rich and stable properties should be powerfully predictive of the features seen in organisms.

Evolutionary theory has grown without insight into the highly self-organized properties of complex regulatory systems. While those properties are yet barely glimpsed, it seems obvious that their proper understanding must deeply enrich our understanding of evolution itself. The preliminary models I have described are not yet a theory, but I hope an indication that an untried pattern of inference may prove fruitful.

ACKNOWLEDGMENTS

It is a pleasure to acknowledge fruitful conversations with Drs. Vahe Bedian, Jo Darken, Warren Ewens, Montgomery Slatkin, Alan Templeton, David Wake, William Wimsatt, and the late Thomas Schopf. This work was partially supported by ACS CD-30, ACS CD-149, NSF PCM 8110601, and NIH GM 22341.

REFERENCES

Abraham, I., and W.W. Doane. 1978. Genetic regulation of tissue-specific expression of *Amylase* structural genes in *Drosophila melanogaster*. *Proceedings of the National Academy of Sciences, USA* 75:4446–4450.

Alberch, P. 1982. Developmental constraints in evolutionary processes. In *Evolution and Development*, ed. J.T. Bonner, pp. 313–332. Berlin: Springer-Verlag.

Aleksander, I. 1973. Random logic nets: Stability and adaptation. *International Journal of Man/Machine Studies* 5:115–131.

Ashburner, M. 1970. Puffing patterns in *Drosophila melanogaster* and related species. In *Developmental Studies on Giant Chromosomes. Results and Problems in Cell Differentiation, Vol. 4*, ed. W. Beerman, pp. 101–151. Berlin: Springer-Verlag.

Axel, R., P. Feigelson, and G. Schultz. 1976. Analysis of the complexity and diversity of mRNA from chicken liver and oviduct. *Cell* 7:247–254.

Bantle, J.A., and W.E. Hahn. 1976. Complexity and characterization of polyadenylated RNA in the mouse brain. *Cell* 8:139–150.

Berge, C. 1962. *The Theory of Graphs and Its Applications*. London: Methuen.

Bishop, J.O. 1974. The gene numbers game. *Cell* 2:81–86.

Bonner, J.T., ed. 1981. *Evolution and Development*. Berlin: Springer-Verlag.

Britten, R.J., and E.H. Davidson. 1969. Gene regulation for higher cells: A theory. *Science* 165:349–357.

Britten, R.J., and E.H. Davidson. 1971. Repetitive and non-repetitive DNA sequences and a speculation on the origins of evolutionary novelty. *Quarterly Review of Biology* 46:111–137.

Brown, D.D. 1981. Gene expression in eukaryotes. *Science* 211:667–674.

Bush, G.L. 1980. In *Essays on Evolution and Speciation*. Cambridge: Cambridge University Press.

Bush, G.L., S.M. Case, A.C. Wilson, and J.L. Patton. 1977. Rapid speciation and chromosomal evolution in mammals. *Proceedings of the National Academy of Sciences, USA* 74:3942–3946.

Cairns, J. 1981. The origin of human cancers. *Nature* 289:353–357.

Charlesworth, B., R. Lande, and M. Slatkin. 1982. A neo-Darwinian commentary on macroevolution. *Evolution* 36:474–498.

Chikaraishi, D.M., S.S. Deeb, and N. Sueoka. 1978. Sequence complexity of nuclear RNAs in adult rat tissues. *Cell* 13:111–120.

Corces, V., A. Pellicer, R. Axel, and M. Meselson. 1981. Integration, transcription, and control of a *Drosophila* heat shock gene in mouse cells. *Proceedings of the National Academy of Sciences, USA* 78:7038–7042.

Davidson, E.H., and R.J. Britten. 1979. Regulation of gene expression: Possible role of repetitive sequences. *Science* 204:1052–1059.

Dickenson, W.J. 1980a. Tissue specificity of enzyme expression regulated by diffusible factors: Evidence in *Drosophila* hybrids. *Science* 207:995–997.

Dickenson, W.J. 1980b. Complex cis-acting regulatory genes demonstrated in *Drosophila* hybrids. *Developmental Genetics* 2:229–240.
Dover, G., S. Brown, E. Cohen, J. Dallas, T. Strachan, and M. Trick. 1982. The dynamics of genome evolution and species differentiation. In *Genome Evolution,* ed. G.A. Dover and R.B. Flavell, pp. 343–372. New York: Academic Press.
Erdos, P., and A. Renyi. 1959. *On the Random Graphs 1, Vol. 6.* Debrecar, Hungary: Institute of Mathematics, University DeBreceniens.
Erdos, P., and A. Renyi. 1960. *On the Evolution of Random Graphs.* Publ. No. 5, Mathematics Institute, Hungarian Academy of Science.
Ernst, S., R.J. Britten, and E.H. Davidson. 1979. Distinct single-copy sequence sets in sea urchin nuclear RNAs. *Proceedings of the National Academy of Sciences, USA* 76:2209–2212.
Errede, B., T.S. Cardillo, F. Sherman, E. Dubois, J. Deschamps, and J.-M. Wiane. 1980. Mating signals control expression of mutations resulting from insertion of a transposable repetitive element adjacent to diverse yeast gene. *Cell* 22:427–436.
Ewens, W.J. 1979. *Mathematical Population Genetics.* Berlin: Springer-Verlag.
Flavell, R. 1982. Sequence amplification, deletion and rearrangement: Major sources of variation during species divergence. In *Genome Evolution,* ed. G.A. Dover and R.B. Flavell, pp. 301–324. New York: Academic Press.
Fogelman-Soulie, F., E. Goles Chaac, and G. Weisbuch. 1982. Specific roles of the different Boolean mappings in random networks. *Bulletin of Mathematical Biology* 44:715–730.
Garcia-Bellido, A. 1975. Genetic control of wing disc development in *Drosophila*. In *Cell Patterning.* Ciba Foundation Symposium 29, ed. P. Porter and J. Rivers, pp. 161–182. Amsterdam: Elsevier-North Holland.
Gillespie, D., L. Donehower, and D. Strayer. 1982. Evolution of primate DNA organization. In *Genome Evolution,* ed. G.A. Dover and R.B. Flavell, pp. 113–134. New York: Academic Press.
Gould, S.J., and N. Eldredge. 1977. Punctuated equilibria: The tempo and mode of evolution reconsidered. *Paleobiology* 3:115–151.
Gould, S.J., and R.C. Lewontin. 1979. The spandrels of San Marco and the Panglossian paradigm: A critique of the adaptationist programme. *Proceedings of the Royal Society (B)* 205:581–598.
Green, M.M. 1980. Transposable elements in *Drosophila* and other diptera. *Annual Review of Genetics* 14:109–120.
Hastie, N.D., and J.O. Bishop. 1976. The expression of three abundance classes of messenger RNA in mouse tissues. *Cell* 9:761–774.
Hough, B.R., M.J. Smith, R.J. Britten, and E.H. Davidson. 1975. Sequence complexity of heterogeneous nuclear RNA in sea urchin embryos. *Cell* 5:291–299.
Humphreys, T. 1963. Chemical dissolution and *in vitro* reconstruction of sponge cell adhesions, Part 1: Isolation and functional demonstration of the components involved. *Developmental Biology* 8:27–47.
Hunkapiller, T., H. Huang, L. Hood, and J. Campbell. 1982. The impact of modern genetics on evolutionary theory. In *Perspectives on Evolution,* ed. R. Milkman, pp. 164–189. Sunderland, Mass.: Sinauer Associates.
Kauffman, S.A. 1969. Metabolic stability and epigenesis in randomly constructed genetic nets. *Journal of Theoretical Biology* 22:437–467.
Kauffman, S.A. 1971. Gene regulation networks: A theory for their global structure and behavior. In *Current Topics in Developmental Biology, Vol. 6,* ed. A. Moscana and A. Monroy, pp. 145–182. New York: Academic Press.

Kauffman, S.A. 1973. Control circuits for determination and transdetermination. *Science* 181:310–318.
Kauffman, S.A. 1974. The large scale structure and dynamics of gene control circuits: An ensemble approach. *Journal of Theoretical Biology* 44:167–182.
Kauffman, S.A. 1983. Developmental constraints: Internal factors in evolution. In *British Society for Developmental Biology Symposium 6: Development and Evolution*, ed. B. Goodwin and N. Holder, pp. 195–225. Cambridge: Cambridge University Press.
Kauffman, S.A. 1984. Emergent properties in random complex automata. *Physica* 10D:145–156.
Kauffman, S.A., R.M. Shymko, and K. Trabert. 1978. Control of sequential compartment formation in *Drosophila*. *Science* 199:259–270.
Kleene, K.C., and T. Humphries. 1977. Similarity of hRNA sequences in blastulae and pluteus stage sea urchin embryos. *Cell* 12:143–155.
Kurtz, D.T. 1981. Hormonal inducibility of rat α_{2u} globulin genes in transfected mouse cells. *Nature* 291:629–631.
Levy, B.W., C.B. Johnson, and B.J. McCarthy. 1977. Nuclear and cytoplasmic RNA complexity in *Drosophila*. In *The Molecular Biology of the Mammalian Genetic Apparatus*, ed. P. Tso, pp. 417–436. Amsterdam: Elsevier-North Holland.
Lewis, E.B. 1978. A gene complex controlling segmentation in *Drosophila*. *Nature* 276:565–570.
Lewontin, R.C. 1974. *The Genetic Basis of Evolutionary Change*. New York: Columbia University Press.
McClintock, B. 1957. Controlling elements and the gene. *Cold Spring Harbor Symposia on Quantitative Biology* 21:197–216.
Moscana, A.A. 1963. Studies on cell aggregation: Decomposition of materials with cell binding activity. *Proceedings of the National Academy of Sciences, USA* 49:742–747.
Paigen, K. 1979. Genetic factors in developmental regulation. In *Physiological Genetics*, ed. J.G. Scandalias, pp. 2–61. New York: Academic Press.
Peterson, P.A. 1981. Diverse expression of controlling element components in maize: Test of a model. *Cold Spring Harbor Symposia on Quantitative Biology* 45:447–456.
Rachootin, S.P., and K.S. Thompson. 1981. Epigenetics, paleontology and evolution. In *Evolution Today: Proceedings of the Second Congress on Systematic and Evolutionary Biology*, ed. G.C. Scudder and J.L. Reveal, pp. 181–193. Pittsburgh: Carnegie-Mellon University Press.
Rendel, J.M. 1967. *Canalization and Gene Control*. New York: Academic Press.
Roeder, G.S., and G.R. Fink. 1980. DNA rearrangements associated with a transposable element in yeast. *Cell* 21:239–249.
Schopf, T.J.M. 1980. In *Paleobiology, Paleoecology and Evolution*, ed. K.J. Niklas, pp. 135–192. New York: Praeger.
Shapiro, R.A. 1981. Changes in gene order and gene expression. *National Cancer Institute Monographs*. In press.
Sherlock, R.A. 1979a. Analysis of the behavior of Kauffman binary networks. I. State space description and the distribution of limit cycle lengths. *Bulletin of Mathematical Biology* 41:687–705.
Sherlock, R.A. 1979b. Analysis of Kauffman binary networks. II. The state cycle fraction for networks of different connectivities. *Bulletin of Mathematical Biology* 41:707–724.
Sherman, F., and C. Helms. 1978. A chromosomal translocation causing overproduction of Iso-2-cytochrome *c* in yeast. *Genetics* 88:689–707.

Simpson, G.G. 1943. *Tempo and Mode in Evolution*. New York: Columbia University Press.
Smith, J.M. 1982. Overview—Unsolved evolutionary problems. In *Genome Evolution*, ed. G.A. Dover and R.B. Flavell, pp. 375–382. New York: Academic Press.
Stearns, S.C. 1982. In *Environmental Adaptation and Evolution: An Empirical and Theoretical Approach*. Stuttgart: Gustav Fisher Verlag.
Struhl, K. 1981. Deletion mapping on a eukaryotic promoter. *Proceedings of the National Academy of Sciences, USA* 78:4461–4465.
Thomas, R. 1979. A Boolean approach to the analysis of complex regulatory systems. *Lecture Notes in Biomathematics*, 29. New York: Springer-Verlag.
Waddington, C.H. 1975. *The Evolution of an Evolutionist*. New York: Cornell University Press.
Wake, D.B., G. Roth, and M.H. Wake. 1983. On the problem of stasis in organismal evolution. *Journal of Theoretical Biology* 101:211–224.
Walker, C.C., and A.E. Gelfand. 1979. A system theoretic approach to the management of complex organizations: Management by exception, priority, and input span in a class of fixed-structure models. *Behavioral Science* 24:112–120.
Webster, C., and B.C. Goodwin. 1982. The origin of species: A structuralist approach. *Journal of Social and Biological Structure* 5:15–42.
Willmer, E.N. 1960. *Cytology and Evolution*. New York: Academic Press.
Wilson, A.C., T.J. White, S.S. Carlson, and L.M. Cherry. 1977. In *Molecular Human Genetics*, ed. R.S. Sparks and D.E. Comings. New York: Academic Press.
Winfree, A.T. 1980. *The Geometry of Biological Time. Vol. 8*. Berlin: Springer-Verlag.
Zubay, G., and D.A. Chambers. 1971. In *Metabolic Pathways*, ed. H.J. Vogel. New York: Academic Press.

II
ECOLOGICAL MODELS

4
How to Be Objective in Community Studies

L. B. SLOBODKIN

What constitutes an ecological community? A swimmer through a reef or walker in a forest knows himself to be in something, but how clear are its borders? Water bodies, islands, and to a lesser degree isolated wood lots are characterized by clear boundaries, at which major alterations in physical conditions may act as barriers to many organisms, but these are leaky to a greater or lesser extent. Active flyers and passive drifting spores and seeds enter and leave. But are there ever perfectly tight boundaries to communities? In nature there is no sealed community other than the biosphere itself (Botkin et al. 1979). Even if there were biologically impervious borders to a region, it is obvious that at least a flux of energy, and probably other materials, would occur across any postulated community border. For most communities, changes of these fluxes would radically alter the capacity for steady-state persistence of the organisms in the community.

If all communities leak at their borders, how great is the integrity of any community? Is a forest more than an aggregation of plants which happen to share a set of environmental needs that the local environment happens to satisfy, as was advocated by the late Professor Whittaker (1953)? Or is it a kind of superorganism as advocated by the school of Clements (1916)? A recent, very careful, analysis of what appeared to be a clear case of discrete community borders seemed to support Whittaker's position, despite the initial supposition that Clements was correct (Ferson, 1985). Ignoring these difficulties forces on us definitions of communities which seem to involve reasonably constant lists of common species over a sufficiently wide area and a sufficiently long stretch of time to be interesting (MacArthur, 1972).

Even after more or less abandoning the problem of exactly defining borders, other almost insurmountable methodological difficulties remain. Complete description of all but the simplest and most aberrant of communities is impossible. Hot springs, dry sand deserts, early primary successional stages on rock surfaces, and areas immediately after volcanic eruptions may have fewer than 10 species; but for most of the earth any reasonably large enclosure might contain several hundred to several thousand species. Any attempt to describe the ecological interactions of all of these will be completely unwieldy. For example, the number of entries in the simplest species interaction matrix will be the square of the number of species present. This means that community ecology never deals with complete descriptions of entire communities, but must of necessity select the kind of information to focus on. Some of the implications of this have been discussed in Slobodkin et al. (1980).

Perhaps the most common simplification introduced by students of ecosystems is the focus on a subset of species. Rather than study the desert community, one studies the desert bird community, or replaces the utter intractability of a coral reef itself by the relative simplicity of only the coral reef fishes or the corals themselves (Loya 1972). While it is easy to argue that choosing a taxonomic subsystem alone is not the best path, it is hard to demonstrate by example how better to proceed. Perhaps focusing on a randomly selected subsample of species may be an appropriate procedure for some purposes. Choosing the taxonomic subcommunity already assumes where the most interesting regulatory processes of a community lie. The whole family of assumptions is embodied in the idea of a "guild" as defined by Root (1967), and in the many ways the term has been redefined through usage.

In short, community ecology suffers from a series of problems. The definition of community may not be clear; the motivations for study of communities may be mixed; and the technical and statistical problems of evaluating the research may be formidable. Given this suite of difficulties, the first question is, Why bother? Obviously there is an intellectual fascination about the subject. Any observer of land or sea is impressed with the web of interactions between diverse organisms in any one place, and a moment's thought raises questions, given the delicacy of all the interactions and the obscurity of most controls, why any but the simplest communities persist. Also, a visit to a rain forest and to a coral reef within one lifetime, let alone within one day, as, for example, in Puerto Rico, immediately raises deeper questions. We know that the enormously diverse coral reef and rain forest organisms have very much the same genetic and biochemical mechanisms, modulated only by evolutionary and ecological circumstances. Do ecological generalizations underlie the differences between communities in the same way that biochemical generalizations underlie the differences between species? Do ecological generalities actually exist?

In the past decade both the reality of the generalizations and the relevance of these supposed basic processes to natural situations have been subjected to serious question. There are, in particular, doubts of the statistical rigor with which they have usually been tested. This has been the focus of extensive and even acrimonious dispute (see the various papers in Strong et al. 1984). The issues are interesting and may be of general importance.

In order to understand the problem, a historical review of some aspects of ecology and of philosophy of science is necessary. After this background material has been presented, I will suggest approaches to ecology that may resolve some of the difficulties.

THE ECOLOGICAL BACKGROUND

There are a few empirical generalizations that have been inferred from field studies. These generalizations include the existence of Eltonian pyramids (Elton 1927), trophic pyramids (Lindeman 1942), and the "HSS hypothesis" (Hairston et al. 1960; Slobodkin et al. 1967). It was also suggested that processes such as inter- and intraspecific competition, as well as various kinds of coevolution and predation, underlay these generalizations, which were mainly based on the mathematical, theoretical, and experimental investigations by a school of remarkably self-confident and imaginative researchers (see Hutchinson 1978). Simple models using these processes and generalizations as a starting point provided insight about field observations, and the exceptions seemed reasonably easy to explain away.

Theoretical mathematical modeling in ecology and evolution has a history of more than a century. Scudo and Zeigler (1978) provide an anthology of theoretical ecology prior to 1941. The work began in earnest during the twenties and was being taught at Yale, Johns Hopkins, and Chicago during the forties. The focus of the present dispute may perhaps be best understood in terms of the past 30 years' history and sociology of ecology. I am neither historian nor sociologist, but I was a participant.

Starting in the early sixties there was an effulgence of speculative mathematical modeling of interspecific competition, encouraged by G. Evelyn Hutchinson and influenced strongly by his students, especially the late Robert MacArthur, and by the curious attitude towards ecological modeling exemplified by Levins (1966), who stated that mathematical models in ecology and evolution could not simultaneously maximize precision, generality, and "reality."

Models were seen as aids to the development of intuition, not necessarily as ways of generating testable predictions. A theoretical model was expected to conform to facts in a general way, but was allowed to leave out large bits of world in the interest of mathematical tractability, and in exchange the modeler would gain in insight. The potential difficulties with this approach, in which esthetics was at least as important as predictive power, were immediately obvious. I pointed out (Slobodkin 1960), that Gause's axiom of interspecific competition (Gause 1934) was untestable. Schoener's (1972) review of MacArthur (1972) listed most of the other theoretical problems inherent in an excessively free-wheeling approach to theory construction.

Nevertheless, the work of MacArthur and his circle was enormously stimulating. The following of Levins's prescription permitted the construction of reasonably quantitative theories which produced fascinating assertions about a nonexistent simulacrum of the ecological world. That is, formality was possible, but the relation to nature was unclear. (The notion of "formalism" will recur in our discussion.) Given the rather loose ground rules for acceptability of theoretical constructs, there was room for imaginative theorizing, which permitted the rapid development of models purporting to predict, within loose limits, how many species would occupy particular islands (MacArthur and Wilson 1967), how niche characteristics should change as a consequence of how many species cooccurred (MacArthur 1972), how various measures of stability should vary with species abundance (May 1975), and so on. The style of optimistic theorizing spread widely. Maynard Smith's initial game-theoretical models required haploidy but were cheerfully used to explain behavior of diploid organisms (Maynard Smith 1974), and Wilsonian sociobiology (Lumsden and Wilson 1981) is replete with models designed to produce "insight" rather than reality or precision. The sixties and seventies saw an explosion of speculative ecology. Much of this work is elegantly summarized in Hutchinson (1978). However, by the late seventies it had become much more difficult to develop charming speculations about ecology. Too many bright young assistant professors and their students had walked over the ground.

Unfortunately, formalism never worked well in biology. Biometry is highly formal but has no biological content in its own right. While the subject matter of arithmetic, and even of certain aspects of physics, is sufficiently circumscribed to permit a reasonable degree of formalism, biological systems must be "simplified" before they can be formalized. Mere narrative or anecdote that fills the pages of Darwin (or Freud, or Marx) is not really in shape for philosophical analysis. If it is stripped of ambiguity and stated in language so pure and mathematics-like that it deserves the name of "formal," a great deal of its rhetorical impact is lost. Formalizations of biology may become so abstruse as to be sterile. Consider Woodger's (1937) attempt to construct a symbolic logic for embryology. While written as a contribution to biology, it is much better known to logicians.

Population dynamics and population genetics contain central theories which are highly formal and mathematically concise, at the cost of requiring extreme simplifying assumptions. Discussions of organisms in natural situations tend to be more verbose. Perhaps discussions of general questions of ecology and evolution cannot be highly formal without simplifying assumptions, but, as we shall see, even the meaning of "assumption" as used in ecology and evolution is often misunderstood.

I have been dubious about this approach for several decades and even complained in print (Slobodkin 1965, 1974a) but without much audience. These complaints were less effective than they ought to have been since they focused on theoretical problems and not on empirical examples. In this sense they were more fulminations than refutations. In addition, many articles had appeared in which the promise of enhanced insight seemed to be fulfilled. For example, the MacArthur memorial volume edited by Cody and Diamond (1975) contained papers on insular distribution which implied that concordant results were forthcoming, despite ostensible methodological flaws. These results were not usually intended to be tests of theory in any rigorous sense, but were investigations of natural history suggested by the theories, during which the investigators became fascinated by biological particularities. As noted by Colwell (1984), community ecology began to divide into subfields focusing on separate processes rather than on description of entire communities. Foraging theory, e.g., became a field in its own right.

THE PHILOSOPHICAL BACKGROUND

Usually, scientists neither know nor care about the philosophy and history of science, any more than a healthy athlete cares about the intricacies of neuromuscular physiology. When movement becomes difficult the situations change. In the last decade Kuhn (1962) on scientific "revolutions" and Popper (1972) on the relation between science and testability not only have been widely read by ecologists but have generated groups of converts. (Conversions of many kinds have been part of the *Zeitgeist* of the past two decades. I suggest that conversion phenomena, particularly among intellectuals, are worthy of serious investigation. Is conversion a consequence of weakness of religious education? Is the previous sentence revolutionary or even testable?)

Kuhn presented historical evidence that science has progressed in a stepwise fashion. For relatively long periods the kinds of questions, methodologies, and answers that are considered interesting in any particular discipline stay the same. Science done during these periods is "normative" or "paradigmatic." Occasionally the meaning of question and answer is itself questioned. Old paradigms are rejected and radically new questions, methodologies, and answers become the center of the discipline. These are the periods of scientific "revolution." Reading Kuhn often stimulates a search for revolution, in almost the same way that one's first reading of Robin Hood sets up dreams of lovable banditry.

Popper's central concern was the role and meaning of testability in science. In brief, he considered scientific discourse superior to nonscientific discourse because of this testability. This assertion is neither invalid nor novel, nor can it be simplistically applied. Just as the long and complex arguments of Darwin and Wallace were taken into the world as a set of mottoes about natural selection resulting in survival of the fittest, Popper's "true believers" have their mottoes. "Science consists of refutation of hypotheses," "If assumptions are not being tested we are talking about tautologies," and so on. Once these are accepted in all their simplicity, it is an easy matter to undertake bellicose destructive criticism of even the most significant biological advances.

The position of Popper in British philosophy of science is that of a late exponent of the fruits of the analytical school of philosophy, characterized by its concern for language and formalism. A monumental work describing this process of formalization for part of mathematics in Whitehead and Russell's *Principia Mathematica* (1910). In one sense the heyday of the analytic school began with the young Wittgenstein (1922) and ended with the older Wittgenstein (1958). However, hope for complete success in the quest for demonstrably complete and consistent formal systems was extinguished, finally, by Gödel (Gödel 1931; Nagel and Newman 1958). Nevertheless, despite Gödel, in a world full of confusion there is unending hope that formalism will clear the air, a hope that seems to be only psychologically, and not philosophically, defensible.

We must keep clear the distinction between the formalism of evaluating particular measurements or data sets and that of organizing the subject matter of the field itself. This distinction may be readily appreciated by considering social sciences (economics, psychology, and sociology) in which masses of data are analyzed with enormous sophistication without permitting the construction of models with clear predictive power.

Neither the work of Popper, nor that of his biologist followers, is "formal" in the same sense as that of Woodger or Whitehead and Russell, or even "theoretical" in the sense of the stepwise construction of explicit theory in population genetics or mathematical population ecology. Some of the arguments arise when critics insist on what they believe are the proper biometric standards applied to other people's data. This insistence itself is commendable, but there is no clear consensus on the meaning of the word "proper." Moreover, much of the polemic arises from attempts to reconcile verbal paraphrases of theory. There is almost always a gap between narrow mathematical theory and the richness of the natural world. Verbalizations cannot extend a formal theory beyond its appropriate domain without losing the precision of the theoretical analysis. This important point deserves clarification.

Many years ago I coined the term "prudent predation" to refer to the choice a predator ought to make among different age categories of prey, if the predator was going to maximize yield per unit of damage to the prey population (Slobodkin 1960, 1974b). The words "prudent predation" were used in the context of a specific equation. They did not mean anything beyond that. In particular, there was no intention to cover the full richness of meaning associated with the literary use of the term "prudence." The conclusion drawn from my analysis was that predators eat what they can get, prey escape as well as they can, and the evolutionary outcome is to make the predator look as if it were "prudent" in the sense defined by the equation. One critic pointed out that proper prudence would require that the predator also could choose the absolute quantity of prey to be taken. Predators that eat everything they can catch should not be called prudent because "the predator is then greedy rather than prudent" (Maiorana 1976). I was invited to write a comment on this criticism. Had I done so, and had others joined the fray, we would have had an ideal example of a simple theoretical concept being described by an English word and then a dispute occurring over the meaning and proper use of English words, as if this constituted a dispute about theory. This is not a unique case by any means. Consider all the wonderful ways in which Wright's concept of an "adaptive landscape" has been used and compare them with the usage by Wright (1931). I am sure most readers can think of other examples.

Obviously, the questioning of empirical or philosophical paradigms and the refusal to slavishly accept traditional doctrines is a healthy sign, but many recently published evolutionary and ecological polemics tend to be so informal as to resemble medieval theological disputes, arguments in courts of law, or discussions between rival schools of psychoanalysis more than they resemble physics or mathematics. The opinions of philosophers of science are often invoked, but it should be recalled that philosophers of science have primarily functioned as analysts, describing scientific history, rather than as authorities on the path science ought to take. The use of philosophy of science as an assay for scientific fineness therefore deserves careful thought. Feyerabend (1975), perhaps in reaction to Popper, has denied, on philosophical grounds, the value of any philosophical criteria for the advance of science. What would it mean if the anarchic recommendations of Feyerabend rather than the strictures of Popper had motivated a methodological reformation? Is there a scientific standard for choosing Popper over Feyerabend? It seems likely that choices between philosophies of science can only be justified historically in terms of the accomplishments of adherents of the rival philosophies. It should also be kept in mind that volume of printed matter can only be equated with scientific accomplishment in the context of tenure and promotion decisions and not in a broader historical context.

Several authors have recently questioned the validity of the accepted understanding of community ecology in an interesting and persistent way. These critics have invoked the authority of Popper, and to a lesser degree Kuhn, but their criticisms are not merely philosophical. These critics have been extremely valuable in promoting more careful empirical analyses. They have, however, gone well beyond that, as we will see.

ASSUMPTIONS AND HYPOTHESES: THE OBJECTIVITY OF TESTS

Having briefly considered the historical and philosophical context I now turn to an analysis of some current arguments about objectivity in community ecology.

Various models have been developed to predict such features as species abundance distributions and island species associations, starting from assumptions about species-species interactions. Simberloff and his associates claimed these predictions had not met proper statistical tests. More recently Connell (1983), among others, claimed that competitive interactions, which generally had been assumed to be of significance in nature, were not of major importance, or at least their importance had not been properly demonstrated in most researched cases. Connell claimed that an "objective" (his word) literature search denied the claim by Schoener (1983) and others that these mechanisms, mainly postulated prior to 1970, had now been adequately demonstrated in nature. The controversy centers on the testing of ecological hypotheses and assumptions in nature.

A focal controversial issue is whether or not "hypotheses" are being properly tested. In the history of science the term "hypothesis" has been successfully used in several different ways. It sometimes is used as in the term "null hypothesis" (see below) and sometimes as in "one of the hypotheses of evolutionary theory is that . . . ." In the second usage, the term hypothesis melts into the term "assumption," and, to further complicate the problem, in biology the word "assumption" is used in at least two contradictory senses. We may "assume" things whose existence is not at all certain, or we may "assume" things that are completely obvious. Mendel did the first, Darwin and Wallace did the second. Confusion between the two is a wonderful starting point for erudite nonsense.

Mendel's breeding experiments gave results which seemed to follow very simple statistical laws. He therefore "assumed" the existence of hereditary factors. He further postulated that these factors were present in pairs in parents and that each parent donated one of each pair at random to its offspring. If the results of subsequent experimentation had been different the eventual conclusion would have been drawn that in fact no such entities existed. The assumption in this case, as in many examples in physics and astronomy, is of the form "Let us imagine that there exists an entity with the properties . . . ." The theoretical arguments and the data are or are not consistent with the assumed entity. It is this kind of theory that is taken as a model by Popper, along with many other philosophers of science, who began by focusing on the history of physics. The theoretical arguments and the data are tests of the validity of the assumptions. The consequences of physical theory are relatively easy to see. They include large explosions, planetary movements, and automobiles. The existence of entities such as electrons, quarks, and pions, for example, is assumed in constructing theories but these entities may not be directly observable.

In contrast, Mendel's contemporaries Darwin and Wallace developed their evolutionary theory by "assuming" the whole pool of data of natural history and zoogeography, as known at the time, as a starting point. In the Mendelian type of theory, there are a small number of clear theoretical steps leading from an undemonstrated set of assumptions to a set of empirical conclusions. Falsification of these conclusions would falsify the assumptions. In the Darwin-Wallace type of theory, the assumptions are known to be empirically valid before theory construction commences, and the steps leading from the assumptions to the new testable conclusions are usually many, and not entirely formal. When the conclusions of this type of theory are falsified, it is not the initial assumptions that are being denied, since they had already been tested before the theory was conceived, but rather the logical procedure of going from these assumptions to the conclusions. Treating these two types of theory as if they were the same thing, or worse, jumping between meanings of the term "assumption" several times during an argument, can lead to confusion, which can hardly avoid generating polemic.

NULL HYPOTHESES

Before the advent of Popperian zealotry, null hypotheses were known to biologists primarily as biometric tools. For whatever reasons, an investigator might want to know if some particular treatment or partitioning of data or relationship between measurements was "significant" in a statistical sense. The significance of an experiment could be assessed by the probability that the experimental results deviated from those that might have been expected had the data conformed to known statistical distributions. If the deviation between results and a null distribution is improbably great, then something "significant" is occurring, without any commitment as to what. These distributions were derived mathematically, not biologically. They were termed "null" distributions, in the sense of being expected if there really was no difference at all that could be assigned to a treatment or to a partitioning of data.

The meaning and usefulness of null hypotheses has become a more serious problem since the development of cheap and rapid computer technology. Prior to this the only null hypothesis distributions available were those calculated from analytically tractable models. Now, Monte Carlo procedures permit computation of distributions from almost any model. In particular, it is possible to generate distributions of numbers from what might be called "biological null models," i.e., from stochastic models which roughly mimic a biological theory, but in which one or more of the biological suppositions has been replaced by an alternative or antithetical one. Data which an author may have taken in support of some theoretical argument, on the basis of a routine biometric significance test, may then be shown to be not significantly different from the expectations of some other null hypothesis, based on more or less biological assumptions. But no data set will survive testing by an infinite battery of null hypotheses. Is it clear which null hypotheses should be used as tests?

Colwell and Winkler (1984) focus on the problem of what constitutes an appropriate null hypothesis. They start by generating a series of hypothetical zoogeographic archipelago distributions from biological null models that either do or do not incorporate competitive interactions of various kinds. They then ask whether or not one could recognize those cases in which interspecific competition had occurred, using the techniques that have been applied in criticism of field data interpretation. They conclude that the construction of "appropriate, unbiased, null models is extremely difficult" (p. 358).

TAUTOLOGY: A BRIEF COMMENT

Derived from early studies by Popper, developed in detail by Peters (1976), and running as a *bas motif* under the discussions of null hypotheses, is the notion that assertions which are derived from a set of assumptions without passing the appropriate null hypothesis test are "tautologies" and that tautologies are empty of scientific content. This gives tautologies a bad name and further, gives the very wrong impression that tautologies are very common, in fact, lurking in every loose argument, waiting to trip up the careless.

Tautologies occur only in formal logical systems, for example arithmetic of integers. Consider the classic tautology that "$1 + 1 = 2$." It can be rewritten as "$1 + 1 = 1 + 1$," "$2 = 2$," or "$2 = 1 + 1$" without any worry that meaning has been deeply changed. In fact the expression in quotation marks has no particular scientific meaning at all, except to the degree that it represents description of, or operations on, objects or processes in nature. Once assigned this representative role the assertion "$1 + 1 = 2$" may lose its tautological character. In fact, it may well be false.

For example, if we want to use the expression "$1 + 1 = 2$" in an empirical context we obviously must ask, "One what?" If the answer is "goose," "bullet," "mass of plutonium," "unicorn," or "cloud," the next question is the operational one, "How can I tell when I have one of these?" and then "Are there actual circumstances when I would have one of these?" All terms in quotation marks can pass the first test of meaningfulness, and only "unicorns" would fail the second one. We must also ask what operations are to be understood by "+" and "=." Are we putting them on a scale together and reading off their weights? Are we putting them on a scale separately, removing them and taking their weights to the nearest integer and then adding those integers, or are we, most simply, counting them after they have been brought into proximity? In the last case, clouds are curious, because two individual clouds brought into immediate proximity result in one cloud, but two clouds of unit weight would have total weight "2." In this operational context "$1 + 1 = 1$" or "$1 + 1 = 2$," depending on what we assign as the meaning of addition. The arithmetic of clouds is not the same as the arithmetic of the mathematicians' integers. Certainly, tautological assertion is possible in meteorology and in biology, but I suggest that the arguments over whether biological theories are or are not tautological often slur over the operations by which the terms are empirically identified. Peters (1976) has a rich collection of what he considers to be tautologies, but they are not. Detailed criticisms of his arguments, case by case, would require too long an exposition (but see Hutchinson 1978, pp. 237–241).

THE LIMITS OF TWO-VALUED LOGIC

Part of the curious misunderstandings about null hypotheses and tautologies has resulted from the common error of believing that the kind of information provided in elementary courses in logic really applies to scientific questions, in particular the mistaken belief that two-valued logic is the way the world is made, or at least the way that science progresses. In two-valued logic, statements are either true or they are not true; we can assert either "a" or "not a" and in most formulations the assertion "not 'not a'" is equivalent to the assertion "a." The usual thinking about statistical testing does not deviate from this. The assertion that a null hypothesis is to be rejected because it has a low probability of being true is carelessly construed as assigning a probability to the absolute truth of the null hypothesis, not as assigning a probabilistic truth to the alternative hypothesis. This distinction may or may not be a serious one. What is much more serious is the unconscious assumption that hypotheses come in pairs, one member of the pair being the null hypothesis, the other member the hypothesis that is being considered by the investigator as being tentatively acceptable if the null hypothesis is rejected. As a rule there is no theoretical reason for identifying rejection of the null hypothesis with acceptance of a single alternative hypothesis.

At any stage of an investigation there are an infinite number of alternative hypotheses, waiting in the wings to replace those that fall victim to empirical tests. Worse still, infinitely large sets of these ghostly hypotheses will serve equally well to describe any particular data set. If this seems an unreasonable assertion, imagine three points on a plane graph. If you like, let them be on a straight line, but it is not necessary for the argument. Now imagine the best line through these points. What did "best" mean in the previous sentence? Did you imagine the best line to be straight? Now add a point slightly above the straight line, so that we have four points. Draw the best curve connecting the four points. Obviously there are an infinite number of such curves that touch all four points and there are also an infinite number of curves that will touch only the first three points and not the fourth. By pure curve fitting we may eliminate enormous numbers of possible alternatives to a null hypothesis, without ever seeming to be restricted to a finite number of still acceptable hypotheses.

Is this nonsense? If so, where did I go wrong and why doesn't the problem actually arise? I suggest that the problem of choosing the wrong member of the set of hypotheses that fit the data equally well usually does not arise, or at least does not persist, for two reasons. First, subsequent data may be expected to prune the set of equally acceptable hypotheses. Second, hypotheses are usually embedded in theories that have multiple sets of consequences. This provides a kind of supplementary test of the reasonableness of a hypothesis. Confronted with two hypotheses which agree equally well with a particular data set we prefer the one that is part of a "richer" theory. We will return to this below.

AN EXAMPLE OF APPLYING TWO-VALUED LOGIC TO LITERATURE SURVEYS

There have recently appeared two papers, one by Connell (1983) and one by Schoener (1983), which are review articles on competition in nature. Both are literature surveys and reviews of recent ecological literature. Both authors counted the number of published tests that show or do not show the existence of competitive interactions in nature. Connell concluded that competition is not really pervasive in nature, and Schoener concluded the opposite. Various authors have joined in favoring one conclusion or the other.

Presumably, scientific literature is clearly written, the body of literature is finite, and the particular papers examined have been refereed. The number of papers reviewed is so large that reading all of them would be out of the question for most investigators. We rely on review articles to tell us the state of current research. Is our reliance misplaced? How can such a disagreement occur, and what does it tell us about the current state of ecology?

The Stony Brook Ecology Reading Group (1986) have painstakingly compared the two review articles to find the source of the difference in conclusions. The two authors approached the literature differently. Schoener attempted to review the literature in what may be called the classical fashion. He sought out as many articles as time and energy permitted, summarized his understanding of whether or not competition had been important in each of the studies, and then presented a statistical analysis of the fraction of studies in which competition mattered. In a sense, he had made himself an expert on the literature in this field and the validity of his conclusions depended on the general quality of his expertise. Connell tried a different way of preparing a review article. For reasons indicated in his paper, he felt that objectivity would be served by setting up a clear sampling domain for his search of the literature. Instead of gathering as much literature as he could, he restricted himself to a small set of journals over a restricted number of years, and within those decided *a priori* how particular kinds of evidence would be scored. The Stony Brook Group demonstrated that this methodological difference is responsible for a large share of the difference in conclusions. They suggest that the apparently more objective method used by Connell may have been influenced by problems of the sociology of science. If competition is accepted dogma at the beginning of a time interval, isn't it more likely that papers will be published purporting to be disproofs rather than those that once again demonstrate the obvious? On the other hand, can we always rely on expertise to be unprejudiced in its use of the literature?

It is probably impossible to specify objective criteria for reviewing scientific literature. It is clear, however, that part of the polemical character of the dispute came from both authors agreeing that the question at issue could be answered by a simple yes or no. Both agreed that competition was either important *or* unimportant. Neither the meaning of "important" nor the importance of the question is completely clear. It is, however, clear that acting as if two-valued logic applied, so that the word "or" must be taken seriously, is a major contributor to misunderstanding.

RICH AND LIMITED THEORIES: CHOOSING BETWEEN TOOLS AND GADGETS

A metaphoric way of thinking about theoretical richness is to consider the distinction between a tool and a gadget. On the streets of most cities there appear, from time to time, sellers of devices specifically designed, for example, to peel grapes. These devices, at least in the hands of the demonstrators, work very well but do nothing else. A gadget is more limited in applications than a tool. A tool, like a fine kitchen knife, might have a slightly more difficult time peeling a grape, but in return is applicable to a whole range of other jobs, all of which it can perform reasonably well. I suggest that some theories have the richness of applicability of a tool while others are more like gadgets.

If two theories are statistically equivalent in their predictions in one overlapping portion of their domains, but one of them can be shown to have a much larger predictive domain, this one is a richer, and therefore preferable, theory. Implicit in this assertion is that there are more predictions from the richer theory and that at least some of these are less subject to statistical ambiguity than those in the region of overlap with the more limited theory. The distinction between "rich" and "gadgety" theories and its possible importance was suggested by an analysis of the evolutionary constraints on *Hydra,* presented in detail elsewhere (Slobodkin et al. 1986).

We can test the richer theory outside of the statistically obscured region where it and the "gadgety" theory share portions of their domain. This deviates from the statistical notion of hypothesis testing, since proper statistical procedure is to choose between hypotheses on the basis of particular data sets. If only one theory offers a hypothesis about a particular data set, the silent theory is not being tested at all. General acceptance of this approach to choosing between theories would stack the deck against theories with more limited domains, and thereby favor more complex theories. Unless the implications of the more complex theories have been very carefully developed, this is likely to lead to temporary acceptance of erroneous hypotheses. There is, however, a converse problem. Consider the possibility that *ad hoc* randomized models designed to fit the zoogeographic data from a single archipelago, and not making predictions about anything else, work very well. In fact, they may fit the data as well as theories based on natural selection and competition. Nevertheless, their domain of prediction may be so narrow that, although they may be useful as gadgets, they cannot be accepted as disproofs of richer theories, unless and until their domains can be expanded in a formal way to match those of their competitors.

This does not mean that theories with narrow domains will never be useful. For example, for three hundred years it has been generally accepted that the heavens are not geocentric. Nevertheless, many navigators of small boats still use a Ptolemaic model of astronomy to determine their locations. Calculations based on a Newtonian system are inconvenient, and the relativistic corrections are completely unnecessary for practical purposes. From the standpoint of what Levins (1966) termed "insight," the heliocentric, relativistic model of the solar system is accepted, while for convenience a much cruder model is preferable.

CONCLUSIONS

Any ecological analysis of a group of organisms is expected to be ultimately connected to community ecology (Colwell 1984), but not necessarily to those aspects of community ecology which have been the focus of polemics. The more descriptive areas of ecology, such as forest ecology, soil ecology, coral reef ecology, limnology, etc., do not seem to generate either acrimonious dispute or philosophical analysis. Community ecology, in the broad sense, however, has problems.

When a research area boils with dispute, it may be a sign of a new scientific revolution, the death of a field, or, most likely, a consequence of misunderstandings. When death of a research area is very close, one finds individual leaders surrounded by small schools of devotees who do not even bother communicating with other schools, and perhaps publish only in their own journals. Post-Freudian psychoanalysis has almost died this way. Think of the tangled interplay among psychoanalysis, clinical psychology, psychiatry, and neurology. If classical psychoanalysis disappeared how much clinical difference would it make? If community ecology disappeared would it matter?

I suspect that the main disputes are where a search for broad but subtle patterns exists for which small-scale mechanisms are invoked as explanations. It is assumed that the patterns are recognizable as statistical properties of entire communities, i.e., extensive variables, and the mechanisms are assumed to be ultimately the intensive variables of evolutionary mechanisms. However, as I have noted elsewhere, "Extensive variables can only in very circumscribed circumstances be considered to be optimized by evolution . . . any theory stated in terms of extensive variables, or even containing extensive variables as a necessary component is to be regarded with suspicion" (Slobodkin 1972). Not following this injunction may be similar to the kind of thinking which derives vague but comforting doctrines of human free will from the Heisenberg uncertainty principle, or explains history through the second law of thermodynamics or through kin selection.

If this is the source of the confusion, then the philosophical arguments in the ecological journals during the past few years have served a valuable function; and if the arguments can now end quietly, community ecology can proceed, with an enhanced awareness of its limitations and its philosophical pitfalls. There exist illegitimately large questions (see Weisskopf 1984). To set the goal of ecology as "understanding communities" may be as useless as setting the goal of physiology as "understanding life." We may have to be grateful for methodologically feasible small bits of that large goal.

The problem of what can be learned about ecological communities is urgent and not merely an intellectual taste. I believe that applied ecology may yet protect ecology from becoming a little, strife-riven academic backwater. The intellectual reservations of ecologists, and their hesitancy to state firm conclusions, do not indefinitely delay the consequences of economic, technological, and demographic events on existing ecological communities. The failure of ecologists to provide answers to applied questions will not mean that "expert" answers will not be supplied by those with sufficient ignorance of ecology who do not see the reasons for reticence (Slobodkin 1985).

What is required is to dissect out intellectually interesting problems, or to follow a clinical approach and focus on applied ecology. However, given the general urgency of ecological questions, it is not legitimate to become embroiled in purely academic or philosophical bickering, or to turn one's back on the field itself.

Obviously, resolving the problems raised in the past several paragraphs would require volumes, even if appropriate information were available, which it is not. I have focused on one rather narrow but absolutely critical question of community ecology: How can we maintain objectivity? Maintaining objectivity does not guarantee correctness or completeness, but rather centers on being clear about our accomplishments. We are confronted with twin dangers. We may either fool ourselves into certainty or worry ourselves into silence. From both an intellectual and practical standpoint it is just as dangerous to fail to draw conclusions as to be satisfied with illegitimate conclusions.

ACKNOWLEDGMENTS

My research has been supported by the National Science Foundation (most recently BSR-8310794), the Biological Science Division of NASA (NAGW 144), the Mobil Oil Foundation, and the Commission of the European Communities (Contract No. ENV-164 I [S]). The manuscript was vastly improved by criticism from Scott Ferson and Rosina Bierbaum.

REFERENCES

Botkin, D.B., B. Maguire III, H.J. Morowitz, and L.B. Slobodkin. 1979. A foundation for ecological theory. *Memorie dell'Instituto Italiano di Idrobiologia* 37(*Suppl.*): 13–31.

Clements, F. 1916. *Plant Succession: Analysis of the Development of Vegetation.* Publication 242, Carnegie Institution of Washington.

Cody, M.L., and J.M. Diamond. 1975. *Ecology and Evolution of Communities.* Cambridge, Mass.: Belknap Press.

Colwell, R.K. 1984. What's new? Community ecology discovers biology. In *A New Ecology: Novel Approaches to Interactive Systems,* ed. P.W. Price, C.N. Slobodchikoff, and W.S. Gaud, pp. 387–396. Chichester: John Wiley-Interscience.

Colwell, R.K., and D.W. Winkler. 1984. A null model for null models in biogeography. In *Ecological Communities: Conceptual Issues and the Evidence,* ed. D.R. Strong, D. Simberloff, L.G. Abele, and A.B. Thistle, pp. 344–359. Princeton: Princeton University Press.

Connell, J.H. 1983. On the prevalence and relative importance of interspecific competition: Evidence from field experiments. *American Naturalist* 122:661–696.

Elton, C.S. 1927. *Animal Ecology*. London: Sidgwick and Johnson.

Ferson, S. An ectone between hemlock and hardwood communities. Does it imply competitive exclusion? Unpublished manuscript.

Feyerabend, P.K. 1975. *Against Method: Outline of an Anarchistic Theory of Knowledge*. London: Humanities Press.

Gause, G.F. 1934. *The Struggle for Existence*. Baltimore: Williams and Wilkins.

Gödel, K. 1931. Über formal unentscheidbäre Sätze der Principia Mathematica und verwandter Systeme I. *Monatshefte für Mathematik und Physik* 38:173–198.

Hairston, N., F.E. Smith, and L.B. Slobodkin. 1960. Community structure, population control and competition. *American Naturalist* 94:421–425.

Hutchinson, G.E. 1978. *An Introduction to Population Ecology*. New Haven: Yale University Press.

Kuhn, T. 1962. *The Structure of Scientific Revolutions*. Chicago: University of Chicago Press.

Levins, R. 1966. The strategy of model building in population biology. *American Scientist* 54:423–431.

Lindeman, R.L. 1942. The trophic-dynamic aspect of ecology. *Ecology* 23:399–418.

Loya, Y. 1972. Community structure and species diversity of hermatypic corals at Eilat, Red Sea. *Marine Biology* 13:100–123.

Lumsden, C.J., and E.O. Wilson. 1981. *Genes, Mind, and Culture: The Coevolutionary Process*. Cambridge, Mass.: Harvard University Press.

MacArthur, R.H. 1972. *Geographical Ecology: Patterns in the Distribution of Species*. New York: Harper and Row.

MacArthur, R.H., and E.O. Wilson. 1967. *The Theory of Island Biogeography*. Princeton: Princeton University Press.

Maiorana, V.C. 1976. Reproductive value, prudent predators, and group selection. *American Naturalist* 110:486–489.

May, R.M. 1975. *Stability and Complexity in Model Ecosystems*. Princeton: Princeton University Press.

Maynard Smith, J. 1974. The theory of games and the evolution of animal conflicts. *Journal of Theoretical Biology* 47:209–221.

Nagel, E., and J.R. Newman. 1958. *Godel's Proof*. New York: New York University Press.

Peters, R.H. 1976. Tautology in evolution and ecology. *American Naturalist* 110:1–12.

Popper, K. 1972. *Objective Knowledge: An Evolutionary Approach*. Oxford: Clarendon Press.

Root, R.B. 1967. The niche exploitation pattern of the Blue-Gray Gnatcatcher. *Ecological Monographs* 37:317–352.

Schoener, T.W. 1972. Mathematical ecology and its place among the sciences. I. The biological domain. *Science* 178:389–394.

Schoener, T.W. 1983. Field experiments on interspecific competition. American Naturalist, 122:240–285.

Scudo, F.M. and J.R. Ziegler. 1978. The golden age of theoretical ecology: 1923–1940. *In:* Levin, S. (ed.) Lecture Notes in Biomathematics, 22. Berlin: Springer-Verlag.

Slobodkin, L.B. 1960. Growth and regulation of animal populations. New York: Holt, Rinehart and Winston.

Slobodkin, L.B. 1965. On the present incompleteness of mathematical ecology. American Scientist, 53:347–357.

Slobodkin, L.B. 1972. On the inconstancy of ecological efficiency and the form of ecological theories. *In:* Deevey, E.S., (ed.) *Growth by intussusception: ecological essays in honor of G. Evelyn Hutchinson.* Transactions of Connecticut Academy of Arts and Sciences, 44:291–307.

Slobodkin, L.B. 1974a. Comments from a biologist to a mathematician. *In:* Levin, S., (ed.) Ecosystem Analysis and Prediction, Proceedings of a Conference on Ecosystems, Alta, Utah, July 1–5, 1974, Siam Institute for Mathematics and Society, pp. 318–329.

Slobodkin, L.B. 1974b. Prudent predation does not require group selection. American Naturalist, 108:665–678.

Slobodkin, L.B. 1979. Problems of ecological description: the adaptive response surface of *Hydra*. Memorie dell 'Istituto Italiano di Idrobiologia, 37 (Suppl.):75–93.

Slobodkin, L.B. 1985. Breakthroughs in ecology. In *The Identification of Progress in Learning,* ed. T. Hagerstrand, pp. 187–196. Cambridge: Cambridge University Press.

Slobodkin, L.B., D.B. Botkin, B. Maguire, Jr., B. Moore III, and H. Morowitz. 1980. On the epistemology of ecosystem analysis. In *Estuarine Perspectives,* ed. V.S. Kennedy, pp. 497–507. New York: Academic Press.

Slobodkin, L.B., F.E. Smith, and N.G. Hairston. 1967. Regulation in terrestrial ecosystems, and the implied balance of nature. *American Naturalist* 101:109–124.

Slobodkin, L.B., K. Dunn, and P. Bossert. 1986. Evolutionary constraints and symbiosis in hydra. In P. Calow, *Physiological Ecology,* Cambridge University Press. In press.

Stony Brook Ecology Reading Group. 1986. Why do Connell and Schoener disagree? *American Naturalist* 127:571–576.

Strong, D.R., Jr., D. Simberloff, L.G. Abele, and A.B. Thistle, eds. 1984. *Ecological Communities: Conceptual Issues and the Evidence.* Princeton: Princeton University Press.

Wallace, A.R. 1858. On the tendency of varieties to depart indefinitely from the original type. In Charles Darwin and A.R. Wallace, On the tendency of species to form varieties; and on the perpetuation of varieties and species by natural means of selection. *Proceedings of the Linnean Society of London* 3:53–62.

Weisskopf, V.F. 1984. The frontiers and limits of science. *Daedalus* 113:177–195.

Whitehead, A.N., and B. Russell. 1910. *Principia Mathematica.* Cambridge: Cambridge University Press.

Whittaker, R.H. 1953. A consideration of climax theory: The climax as a population and pattern. *Ecological Monographs* 23:41–78.

Wittgenstein, L. 1922. *Tractatus Logico-Philosophicus.* London: Routledge and Kegan Paul.

Wittgenstein, L. 1958. *Philosophical Investigations.* Translated by G.E.M. Anscombe. Oxford: Blackwell.

Woodger, J.H. 1937. *The Axiomatic Method in Biology.* Cambridge: Cambridge University Press.

Wright, S. 1931. Evolution in Mendelian populations. *Genetics* 16:97–159.

5

On the Use of Null Hypotheses in Biogeography

PAUL H. HARVEY

The requested title of this chapter was "Neutral Models in Biogeography." That topic is vast and for didactic reasons I shall focus on just one area of current debate. That area could have been one of many (Harvey et al. 1983; Strong et al. 1984). What limits the distributions of animals and plants in nature? Are there regularities in the statistical patterns of species abundance curves? How has MacArthur and Wilson's theory of island biogeography fared over the years? How does the number of species change with island area and why? How are species distributions among islands influenced by competitive exclusion? How common is character displacement? I have chosen to focus on the fifth problem, the detection of the effects of interspecific competition on islands, because it has probably been the most contentious issue and because it so clearly illustrates the difficulty of defining appropriate null hypotheses in ecology. In the context of the work described in this chapter, "neutral models" are synonymous with statistical null hypotheses.

Ecologists have long been fascinated by the extent to which interspecific competition influences the geographical distributions of species. This seemingly straightforward question cannot be easily answered. On first encounter, species occurrences on the islands of archipelagoes offer suitable material with which to work, particularly if the study organisms are plants or vertebrates with poor dispersal abilities: Not all species are found on each island, habitats can be impoverished, population sizes are often small, and islands may be isolated from each other. As a consequence, the effects of competitive exclusion might be particularly evident. However, scores of published studies have applied a variety of analytical techniques to such data sets but no consistent patterns appear to have emerged. Indeed, the conclusions drawn by different authors working on the same data set are often at variance because they have used different assumptions when designing their statistical models (several examples are reviewed in Harvey et al. 1983). Even when patterns can be interpreted as being caused by interspecific competition, other explanations are also possible. Evidence of competition from species distributions will always be circumstantial, though, of course, some patterns are stronger or more convincing than others. Ultimately, manipulative field experiments are always desirable, although these are often not practical.

This topic has been reviewed several times recently, though in variable depth and from a variety of perspectives (several chapters in Strong et al. 1984). My aim here is to summarize the salient arguments together with illustrative examples. I start by comparing the different ways in which tables showing the presence versus absence of different species on an archipelago have been analyzed. I then point to the problems inherent in such analyses, partially illustrating them with the results of a recently published simulation model of Colwell and Winkler (1984). The same results also provide guidelines for the design of more appropriate statistical null hypotheses against which to test the available data. Improved methods are evident in several recent papers, and, by way of example, I shall describe one of those, by Graves and Gotelli (1983).

A TAXONOMY OF ANALYSES OF SPECIES × ISLAND TABLES

Presented with a species × island table, how should we seek out evidence of competitive exclusion? A common technique is repeatedly to rearrange the ones (presence of species) and zeros (absence of species) in the table, and then compare these random "null tables" with the original data. Sometimes analytical expectations can be derived, but repeated simulations can also be carried out in order to find the distribution of various statistics associated with the null table. However, using these techniques to derive expectations, we might lose from our table some pattern that was not caused by competition. In attempts to conserve such pattern in the null tables, three common restrictions are employed. These are listed below with three common statistical techniques used to detect the presumed effects of competition.

Restriction on the Analysis

Retain number of species per island. Some islands are larger or nearer source pools than others and, for those reasons, might be expected to contain more species. If we rearrange the data without retaining the same number of species on each island, then we may have created differences between our null and observed distributions that did not result from interspecific competition. When creating null tables, it may be useful to rearrange the cells, subject to the constraint that the number of species on each island is conserved.

Retain relative abundance of species. Some species are more common than others, and this need not be caused by competition. For example, smaller species occupy smaller territories and have higher population densities than larger species. We may wish to conserve the relative species abundances in our rearranged table.

Retain incidence functions. Some species are found only on islands of a particular size. For example, among birds, "super-tramps" (Diamond 1975) are found only on small islands. We can conserve this aspect of the original table, that is, species incidence functions, by constraining the rearranged table so that particular species are found only on islands of the sizes on which those species were actually recorded.

Statistical Models Used

We must now decide how to detect differences between the original and the rearranged data. Three methods of analysis have commonly been used.

The number of missing species combinations. Are significantly more pairs of species missing than in the null tables? This is a commonly used but very weak statistic. For example, a species pair might reasonably be expected to occur on 100 islands of an archipelago but only occur on one island; the species pair would not be recorded as a missing species combination.

Comparing species distributions. How often do particular species combinations occur on islands in the null tables compared with the actual data? If species pairs occur less often than expected, they have complementary species distributions, possibly as a result of competitive exclusion.

Comparing islands. Are some islands more or less alike in terms of species composition than the null tables might lead us to expect? Some islands may be dissimilar because particular combinations of species are able to competitively exclude others.

Combinations of Restrictions and Statistical Models

I have outlined three possible types of restrictions that may be used in any combination for constructing the null tables, which makes seven types of null tables (restrictions 1, 2, 3, 1 & 2, 1 & 3, 2 & 3, or 1 & 2 & 3). I have also mentioned three types of statistical tests that have been used to compare the null tables with the observed. Any one of the three tests could be used in combination with any of the seven types of constraints, which gives up to 21 ways the data might be analyzed. At least 11 of those combinations have been used in published studies, and even for subsets of the Galapagos avifauna seven different combinations have been used (see Harvey et al. 1983). It is hardly surprising that different conclusions have been drawn by different investigators. It should be emphasized that Simberloff, Strong, and others working from Florida State University have not only been influential in promoting the above techniques but they have also been careful to emphasize differences among them (see Strong et al. 1984).

PROBLEMS WITH ANALYSES OF SPECIES × ISLAND TABLES

Colwell and Winkler's Simulation

A recently published simulation study by Colwell and Winkler (1984) sheds light on a number of the problems associated with the types of analyses outlined above.

Their technique is to generate imaginary archipelagoes, each populated according to specific rules. Using the worlds they have created, they explore the extent to which different methods of analysis can detect the extent to which competition has actually governed the distribution of species in their archipelagoes.

They begin by generating phylogenies according to a simple simulation model designed by Raup and Gould (1974). As the hypothetical species evolve, their phenotypes change gradually so that, on average, closely related species are more like each other than are more distantly related species. For the purposes of the model, easily measurable characters such as beak length and width are chosen. This evolution takes place on a large land mass under the guidance of a computer routine called GOD. The organisms then disperse to the islands of a nearby archipelago, their destinations being determined by additional computer routines summarized in Figure 5.1.

One routine allows random invasion of the different islands, with subsequent interspecific competition determining which species establish successful populations on each island. When phenotypically similar species invade an island, some

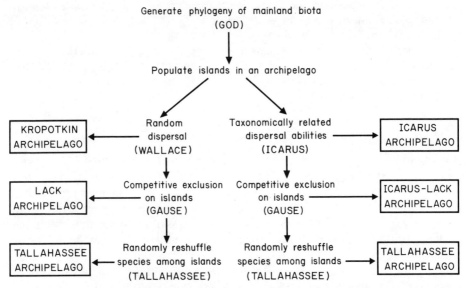

Figure 5.1. Colwell and Winkler (1984) generate a computer phylogeny using a routine called GOD, and distribute species among the islands of an archipelago according to specified rules using either WALLACE or ICARUS. The species on each island are then subjected to competition using GAUSE, thus producing archipelagoes with new species compositions. Finally, the species on the postcompetition archipelagoes are redistributed among the islands using TALLAHASSEE. Recent analyses of real data sets search for the effects of competition by comparing the equivalent of the LACK and ICARUS-LACK ARCHIPELAGOES with the TALLAHASSEE and ICARUS-TALLAHASSEE ARCHIPELAGOES, respectively. However, the comparisons consistently underestimate the effects of competition. More suitable null models against which to compare the LACK and ICARUS-LACK ARCHIPELAGOES are the KROPOTKIN and ICARUS ARCHIPELAGOES, respectively.

are more likely to become locally extinct as a result of competitive exclusion. An alternative routine allows random invasion of islands from the mainland without subsequent competition. As competition occurs only in the first routine, the second makes a suitable null model for comparison.

But a different type of comparison is often made with real data. Species found on different islands are randomly redistributed among the islands (by computer of course); the reassortment is then compared with the original distribution pattern (see Strong et al. 1984). Colwell and Winkler incorporate a computer routine that mimics this procedure. They find that the species distributions differ very little between the reassorted archipelago and the one on which competition has occurred, thus confirming that the effects of competition are indeed partly incorporated into the null model derived from the reassortment routine. Consequently, a comparison that uses this procedure is less likely to detect the effects of competition than is Colwell and Winkler's first procedure (see Grant and Abbott 1980; Case and Sidell 1983).

Another factor, usually ignored, that leads to underestimating the effects of competition, is the varied abilities of different taxa to disperse (see Simberloff 1978; Grant and Abbott 1980). For example, kingfishers are more vagile than kiwis, and so would be likely to be found on islands that kiwis have not yet reached. In a second series of comparisons, Colwell and Winkler allow dispersal abilities as well as morphological characteristics to evolve in the phylogenies generated by their computer. The result is that closely related species tend to end up on the same islands. Some species are extirpated when competition occurs, thus producing a species assemblage on each island that is taxonomically less biased than that of the original invaders. Because differential dispersal abilities produce the opposite pattern from competition, if they are ignored when designing null models the effects of competition will be consistently underestimated.

Colwell and Winkler's model assumes that competition is more intense between closely related than distantly related species. This is also likely to be the case in nature. Since the computer has a complete phylogenetic record, it can assign relative taxonomic status with accuracy. Comparisons made among closely related taxa, for example among species within a genus, are more effective for detecting the statistical consequences of competition (for they represent the level at which the process acts most strongly) than those that incorporate a wider taxonomic spread, such as an order or class. Here, the creator (GOD, or Colwell and Winkler if you will) has a considerable advantage in knowing the likely relations between competitive intensity and the phylogenetic relationship among species, whereas such relations may be far less clear in the real world. Colwell and Winkler could, if they wished, give objective and usable criteria for assigning subsets of species into "guilds" (species with similar patterns of resource utilization; Root 1967), whereas such decisions about grouping real species usually involve some subjectivity.

Using Colwell and Winkler's example, I have introduced just three of the problems associated with the interpretation of species × island tables. They are (1) incorporating the effects of competition into the null model, (2) ignoring the differential dispersal abilities of different species, and (3) the need to take into account the likely importance of competition between species. Other authors have, of course, fallen into or mentioned these and other pitfalls. I give some examples below.

Poorly Conceived Data and Null Tables

It may be that a table recording all the bird species on the islands of an archipelago is correct, but that does not mean that the same table provides a useful data base from which to look for the effects of competition. Each of Colwell and Winkler's three problems is evident. First, competition may have extirpated some species from all the islands; only knowledge of the appropriate source pool can alleviate this problem. Colwell and Winkler's study assumes migration from a nearby landmass rather than speciation with adaptive radiation on the archipelago such as occurred on the Galapagos and Hawaiian islands. In the latter cases the source pool problem can only be dealt with by reference to the fossil record. Second, different dispersal abilities may lead to negative associations between pairs of species across islands; knowledge of dispersal abilities is necessary. Third, several recent discussions have focused on the fact that tables should only contain those species that are likely to be in competition (Alatalo 1982; Diamond and Gilpin 1982; Wright and Biehl 1982). Classic analyses in the literature have used tables that include owls with hummingbirds and ducks with warblers (see Connor and Simberloff 1978); these species pairs may be found in similar habitats but they are not likely to compete for resources.

One other important reason why null tables are often ill-conceived is that some islands do not contain certain habitat types. As a consequence, species that are restricted to particular habitats would not be found on all islands, irrespective of interspecific competition. For example, marsh-nesting birds will not be expected to occur on islands without marshes. A timely call for species' autecologies to be considered when defining island source pools has been made in a recent paper by Graves and Gotelli (1983), to which I shall return below.

Weak Statistical Tests

Once the data and null tables have been well defined, it is necessary to compare them. Weak statistical tests are undesirable because type II statistical error is high (the failure to reject an incorrect null hypothesis). Such error can be increased in a number of ways. When observed data are used to calculate expected cell frequencies, as more constraints (see above) are added to the analysis then the observed data increasingly approximate the expected. As has been pointed out elsewhere, if "(i) the number of islands remains equal to the actual number; (ii) the number of species remains equal to the actual number; (iii) each species is present on exactly as many islands as in actuality; and (iv) each island has as many species as does the real island" (May 1984), in the limiting case of just two species with mutually exclusive distributions across islands, the observed and null tables will have statistically indistinguishable checkerboard patterns (Gilpin and Diamond 1984).

Some statistical tests are stronger than others. I have already mentioned that the number of missing species combinations compared with that expected from a null model provides a much weaker statistic than comparing species distributions across islands. It is clearly desirable to compare full distributions rather than summary statistics whenever possible. A clear example of such a summary statistic shielding strong patterns is provided by Connor and Simberloff's (1979) test that examines all pairwise species combinations across islands and counts how many of them are found on various numbers of islands—33 species pairs on no islands, 14 species pairs on one island, and so on. This distribution is then compared with a simulated distribution using a chi-squared test. Although the two distributions may not differ significantly, individual cases in the observed distribution may be highly unlikely. In the Bismarck Archipelago, 1,690 pairs of bird species share no island, compared with an expected 1,388 pairs according to Diamond and Gilpin (1982). Although this difference is formally significant, it "greatly understates the case [because] buried in this total are pairs of ecologically close species with exclusive distributions despite each species occupying many Bismarck islands, such that the probability of attaining such a result by chance is as low as 6×10^{-9}" (Diamond and Gilpin 1982).

Not only have weak statistical tests been fairly commonplace in this literature, but on occasion the statistics themselves have been incorrectly calculated (e.g., as a result of too few iterations when rearranging the values in actual tables to form null tables) or interpreted (e.g., observed values lying 4.7 standard deviations from the expected mean have been accepted as in accord with the null hypothesis). These types of error have been reviewed elsewhere (Harvey et al. 1983).

THE INTERPRETATION OF RESULTS

If a successful statistical analysis can be completed, taking into account all of the above problems, then groups of species with positively and negatively associated distribution patterns will be identified. One of the most thorough analyses that has been completed to date is by Gilpin and Diamond on the avifauna of the Bismarck Archipelago (1982). Despite various statistical shortcomings that have been discussed elsewhere (Connor and Simberloff 1984), the study does demonstrate the difficulties associated with interpreting significant results. Gilpin and Diamond found far more species with positively than negatively associated distribution patterns. The positive associations were considered to result from shared habitats, endemism (speciation tends to occur on large islands where extinction rates are low), shared geographical origins, and shared incidence functions. On the other hand, negative associations were thought to result from competitive exclusion, differing incidence functions, and differing geographical origins. We might add unequal distributions of habitats across islands as an additional factor contributing to negative species distributions. Diamond and Gilpin used other forms of evidence in order to distinguish among these explanations in particular cases. However, there is always the problem that a force selecting for positive association (e.g., shared geographical origins) may be balanced by one selecting for negative association (e.g., competitive exclusion). Unless all the constraints are built into the original statistical model, cases of competitive exclusion may be overlooked because the null hypothesis accords with the observed distribution.

NEOTROPICAL LAND-BRIDGE AVIFAUNAS: A CASE STUDY

The recent shift of emphasis towards a statistical approach to distributional data by Grant, Simberloff, Strong, and their colleagues has lent improved rigor to ecology. A good example is provided by comparing Graves and Gotelli's (1983) study of the bird species occupying various neotropical land-bridge islands with previous investigations. Graves and Gotelli set out to answer two questions:

(i) At the family level, are island communities a non-random subset of adjacent mainland communities?

(ii) Are species with restricted mainland distributions under-represented on land-bridge islands?

Graves and Gotelli chose seven neotropical land-bridge islands off the north coast of South America and the south coast of the Panama peninsula (Coiba, San Jose, Rey, Aruba, Margarita, Trinidad, and Tobago). How did they overcome problems ignored or inadequately dealt with by previous investigators?

Defining total source pools. Defining source pools has been a continuing problem. Most investigators have defined a single source pool for all the islands studied. We have already seen that this led to problems with Gilpin and Diamond's analysis of the Bismarck data where positive and negative associations could be attributed to shared or different geographical origins—that is, the same or different source pools. Another example is given by using a single source pool for all the islands of the West Indies: Both the Zapata wren (*Ferminia cerverai*) and St. Vincent's parrot (*Amazona guildingii*) are included in the source pool for Grenada, yet "the distance from Grenada to St. Vincent is only 145 km, whereas the Zapata Swamp (Cuba) is over 2400 km away" (Graves and Gotelli 1983). Ideally, then, a different source pool should be used for each island.

Source pools transcend political boundaries. Several authors have used species lists for selected countries to define an appropriate species pool, a practice which should be avoided whenever possible.

Graves and Gotelli's largest island was Trinidad. They found that a source pool including species found within 300 km of the island was appropriate on the grounds that the source pool contains the species found on Trinidad, but that going beyond that distance included many habitats not found on Trinidad. Thus the "total source pool" for each of their islands was designated as all the bird species found within 300 km.

Defining habitat source pools. Birds occur only within a certain range of habitats, which differ among species. Only certain habitats are found on particular islands and, taking into account species autecology, Graves and Gotelli narrowed their total source pools for islands to "habitat source pools" containing only those species occurring in the actual habitats found on particular islands.

Defining colonization potential. Graves and Gotelli believe that incidence functions may be less useful than total geographic range for measuring different species island colonization potentials. Species with wide geographical ranges were considered to be likely to have higher colonization potential than those with restricted ranges. Using various measures of geographical range, they were able to answer question (ii) above.

In summary, using appropriate statistical techniques, Graves and Gotelli were able to show that (1) certain families were consistently overrepresented on the islands (pigeons, flycatchers, and warblers), whereas no family was consistently underrepresented; (2) the habitat pool is a significantly better predictor of species number in each family than the total pool; (3) the number of families on each island is consistent with the expected value; and (4) species with widespread mainland ranges are disproportionately overrepresented on several islands.

Of course, Graves and Gotelli's study is not perfect. Even the casual reader could focus on unanswered questions or methods that could be improved, but such is the nature of ecology. Nevertheless, their study focuses our attention towards species autecologies, and perhaps this is a useful pointer. For example, whether their preferred habitat is present may be of paramount importance in determining whether vagile animals like birds even attempt to breed in an area. The success of breeding attempts may be more dependent on the outcome of interspecific competition for resources.

CONCLUDING REMARKS

Species distributions may be influenced by a number of physical and biotic factors: climate, nest sites, food, dispersal abilities, competition, predation, and parasitism, to name a few. Furthermore, interactions between those factors also come into play. This is one reason why ecology is such a difficult subject from which to get precise answers. Data on the geographical distributions of species must be carefully analyzed with meticulous attention being paid to the design of appropriate null hypotheses. However, such analyses merely provide information that complements more detailed studies on species autecologies, manipulative field and laboratory experiments, and behavioral as well as genetic analyses. Taken together, Grant's studies of Darwin's finches, Roughgarden's and Schoener's on lizards, and Strong's and Lawton's on phytophagous insects (references to all of which can be found in Strong et al. 1984) demonstrate that, as May (1984) has emphasized, there is no One Way to make progress in ecology.

REFERENCES

Alatalo, R.V. 1982. Bird species distributions in the Galapagos and other archipelagoes: Competition or chance? *Ecology* 63:881–887.

Case, T.J., and R. Sidell. 1983. Pattern and chance in the structure of model and natural communities. *Evolution* 37:832–849.

Colwell, R.K., and D. Winkler. 1984. A null model for null models in biogeography. In *Ecological Communities: Conceptual Issues and the Evidence,* ed. D.R. Strong, D. Simberloff, L.G. Abele, and A.B. Thistle, pp. 344–359. Princeton: Princeton University Press.

Connor, E.F., and D. Simberloff. 1978. Species number and compositional similarity of the Galapagos flora and avifauna. *Ecological Monographs* 48:219–248.

Connor, E.F., and D. Simberloff. 1979. The assembly of species communities: Chance or competition? *Ecology* 60:1132–1140.

Connor, E.F., and D. Simberloff. 1984. Neutral models of species' co-occurrence patterns. In *Ecological Communities: Conceptual Issues and the Evidence,* ed. D.R. Strong, D. Simberloff, L.G. Abele, and A.B. Thistle, pp. 316–331. Princeton: Princeton University Press.

Diamond, J.M. 1975. Assembly of species communities. In *Ecology and Evolution in Communities,* ed. M.L. Cody and J.M. Diamond, pp. 342–445. Cambridge, Mass.: Harvard University Press.

Diamond, J.M., and M.E. Gilpin. 1982. Examination of the "null" model of Connor and Simberloff for species co-occurrences on islands. *Oecologia* 52:64–74.

Gilpin, M.E., and J.M. Diamond. 1982. Factors contributing to non-randomness of species co-occurrences on islands. *Oecologia* 52:75–84.

Gilpin, M.E., and J.M. Diamond. 1984. Are species co-occurrences on islands non-random, and are null hypotheses useful in community ecology? In *Ecological Communities: Conceptual Issues and the Evidence,* ed. D.R. Strong, D. Simberloff, L.G. Abele, and A.B. Thistle, pp. 297–315. Princeton: Princeton University Press.

Grant, P.R., and I. Abbott. 1980. Interspecific competition, island biogeography and null hypotheses. *Evolution* 34:332–341.

Graves, G.R., and N.J. Gotelli. 1983. Neotropical land-bridge avifaunas: New approaches to null hypotheses in biogeography. *Oikos* 41:322–333.

Harvey, P.H., R.K. Colwell, J.W. Silvertown, and R.M. May. 1983. Null models in ecology. *Annual Reviews of Ecology and Systematics* 14:189–211.

May, R.M. 1984. An overview: Real and apparent patterns in community structure. In *Ecological Communities: Conceptual Issues and the Evidence,* ed. D.R. Strong, D. Simberloff, L.G. Abele, and A.B. Thistle, pp. 3–16. Princeton: Princeton University Press.

Raup, D.M., and S.J. Gould. 1974. Stochastic simulation and evolution of morphology—Towards a nomothetic paleontology. *Systematic Zoology* 23:305–332.

Root, R.B. 1967. The niche exploitation pattern of the Blue-Gray Gnatcatcher. *Ecological Monographs* 37:317–352.

Simberloff, D. 1978. Using island biogeographic distributions to determine if colonization is stochastic. *American Naturalist* 112:713–726.

Strong, D.R., D. Simberloff, L.G. Abele, and A.B. Thistle, eds. 1984. *Ecological Communities: Conceptual Issues and the Evidence.* Princeton: Princeton University Press.

Wright, S.J., and C.C. Biehl. 1982. Island biogeographic distributions: Testing for random, regular and aggregated patterns of species occurrence. *American Naturalist* 119:345–357.

III
PALEONTOLOGICAL MODELS

6

Neutral Models in Paleobiology

DAVID M. RAUP

Neutral models are useful in testing hypotheses about process. In the typical case, a pattern is seen in empirical data and a mechanism is proposed to explain the pattern. A neutral, or null model is then constructed to answer the question, Would the same pattern have occurred in the absence of the proposed mechanism? The neutral model is tested either through simulation or direct computation. If the pattern produced by the neutral model is indistinguishable statistically from the real world pattern, there is no compelling reason to accept the proposed explanation. If this match is not found, the proposed explanation is not proven but its credibility is at least sustained.

Neutral models are of great value with paleobiological data and often are the only available way to test hypotheses. In this chapter, a variety of applications will be described by summarizing cases associated with the following questions: (1) Did the trilobites go extinct through bad luck or bad genes? (2) Are the major extinctions in the fossil record periodic? (3) Do Ordovician faunas illustrate dynamic equilibrium in diversity? (4) Is the patchiness of the evolutionary filling of morphospace a stochastic or deterministic phenomenon?

The neutral models used in paleobiology do not differ in any substantial way from the null hypotheses used standardly in applied statistics. Consider the classic case of coin flipping. If a coin is flipped 100 times and comes up heads only 30 times, one could propose a number of explanations involving asymmetry of the coin, peculiarities of the flipping process, and so on. The conventional procedure in this case would be to test the observed results against a null hypothesis of total randomness using chi-squared or other techniques. The null hypothesis assumes that none of the proposed explanations is correct and only if this hypothesis can be rejected are any of the proposed explanations viable. The neutral models described in this chapter differ from the coin-flipping case only in being more precisely tailored to the biological problems at hand.

EXTINCTION OF THE TRILOBITES

The trilobites were a highly successful biologic group through much of Paleozoic time. About 75% of all fossil species described from Cambrian rocks are trilobites, and although the vagaries of sampling and preservation make estimates of actual numbers difficult, there were probably about 6,000 trilobite species living at any given time in the Cambrian (Valentine et al. 1978; Raup 1981). This means a standing diversity (species richness) somewhat greater than for mammals or echinoderms today. As indicated in Figure 6.1, trilobite numbers dwindled steadily through Paleozoic time and the group died out completely in the Permian. The question is, Why?

It has been conventional to assume (or presume) that the trilobites "did something wrong." They may have been unable to compete with evolving fishes or they may have been affected by other invertebrate groups. Or changes in the physical environment may have caused the extinction. A number of plausible scenarios have been presented but proof of any of them is virtually impossible. As an alternative, it has been claimed that the decline and ultimate extinction of the trilobites could simply have been the result of bad luck: An accidental excess of species extinctions over originations so persistent that standing diversity dropped to zero (Raup et al. 1973). This is analogous to the possibility of a gambler's failure even in a fair game. The question for the trilobites becomes one of bad genes or bad luck.

We know, of course, that there must have been many more species extinctions than orginations among trilobites after the Cambrian; otherwise the group would have survived. The question is whether this excess of extinctions is statistically plausible.

The neutral model in this case is one that envisions trilobite cladogenesis as

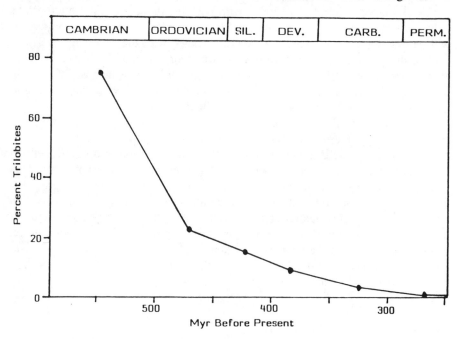

Figure 6.1. Trilobite species diversity through geologic time as a percentage of the total invertebrate fossil record. (Data from Raup 1976.)

a purely random process with average rates of speciation and extinction no different from the averages for other organisms living in Paleozoic seas. If the extinction of the trilobites is credible under these hypothetical conditions, then there is no reason to claim that the trilobites did anything wrong. Alternatively, if extinction of a group as large as the trilobites is not credible under the conditions of the neutral model, then "bad luck" in the sense of statistical accident is not the explanation.

A random birth-death model was applied to this problem by Raup et al. (1973) with the conclusion that a diversity pattern like that seen in Paleozoic trilobites occurs commonly just by chance. The analysis was done by Monte Carlo simulation and suffered from improper scaling, as was later pointed out by Stanley et al. (1981), so the conclusion of the original study was not robust. A more rigorous analysis was performed later using equations that compute the probability of the extinction of a group within a specified time interval, given estimates of species extinction and origination probabilities, and a starting diversity (Raup 1981).

The more rigorous analysis showed that if trilobites had speciation and extinction probabilities typical of other Paleozoic invertebrates, the probability that the whole group would have gone extinct simply by bad luck is vanishingly small! With a starting diversity of 6,000 species in mid-Cambrian, and with the remaining 350 million years of the Paleozoic available, the probability of total extinction turned out to be only 10^{-82}. Similarly low probabilities were obtained when the estimate of starting diversity was lowered to 1,000. Only if that diversity is reduced to 100 or less does the "bad luck" explanation become credible and this explains the misleading results of the Monte Carlo studies.

The neutral model can be rejected with confidence, thus opening the way for deterministic explanations of the trilobite extinction. This does not tell us, of course, what the trilobites did wrong (or what the other organisms may have done better) but it does justify a search for causes. The importance of the exercise is to provide a rigorous test of a subjective and largely intuitive judgment about the causes of a pattern observed in real data.

PERIODICITY OF EXTINCTION

Several studies have concluded that the major extinction events of the past 250 million years of geologic time are uniformly spaced (most recently: Fischer and Arthur 1977; Raup and Sepkoski 1984, 1986; Rampino and Stothers 1984; Sepkoski and Raup 1986). Indeed, one's subjective impression of the data is that the spacing between pulses of extinction is nonrandom, especially because truly random distributions tend to show marked (and largely counterintuitive) "clumpiness" with long gaps between clumps. But the proposition of periodicity must be tested. What follows is an abbreviated account of a complex analysis (Raup and Sepkoski 1984; Sepkoski and Raup, 1986).

The neutral model in this case is one that assumes the extinctions to be randomly distributed in time, and the question is whether a random process will produce patterns of events as periodic as observed in the real data. If the observed level of periodicity is within the reasonable expectations of the neutral model, significant periodicity for the real data cannot be claimed.

There are many ways in which the neutral model could be tested, and the choice depends on a number of special characteristics of the data and some judgments. Of greatest importance is the limited time resolution of extinction data: in the data set for marine families (Sepkoski 1982), times of extinctions are known only at the level of the stratigraphic stage. The stages during the last 250 million years average about six million years. Thus, the data are forced into about 40 "bins." Furthermore, extinction pulses cannot be separated if they occur in adjacent stages. This means that there is a minimum spacing of about 12 million years between extinction events (the Nyquist limit). A much larger data set for marine invertebrate genera has somewhat better time resolution but the binning problem is still significant (Raup and Sepkoski 1986).

Because of the binning problem, the distribution of events in time is very different from what would be produced by a pure Poisson process. If points are dropped at random on a time line, they can be (and often are) very close together. Clearly, it would be nonsensical to test the constrained distribution of the real data against an unconstrained Poisson-produced distribution. This illustrates a common problem in formulating neutral models: The model is meaningful only if it reproduces all important aspects of the real data except those which one wishes to neutralize.

The problem can be approached by using a randomization technique whereby the neutral model is implemented by scrambling the temporal order of the real data. In the case at hand, 12 stratigraphic stages (bins) were identified in the family data as having elevated extinction intensity. The pattern seemed periodic because of the relatively constant intervals between events. The question for testing is whether the uniformity of spacing would be expected by chance in a constrained system such as this. The expectations of chance are evaluated by analysis of repeated randomized versions of the real time series. Randomization is accomplished by shuffling the stratigraphic time scale (the stages with their durations) and then dropping 12 events on the new time scale. The process is constrained to prevent events falling in adjacent stages. For each of a large number of these simulations, periodicity is tested with the metric that was used for the real data. This yields a probability distribution for periodicity under the terms of the neutral model.

The foregoing procedure has shown that the periodicity observed in the real data has an extremely low probability of occurring by chance and the null hypothesis of randomness can be rejected.

A comment should be added about the choice of the 12 extinction events from the total extinction data set (see Fig. 6.2). The 12 were identified as local maxima in the extinction record of families, whereas only eight of these are significantly above background levels (Sepkoski and Raup 1986; Raup and Sepkoski 1986). The question is, Which set of extinctions should be used? It turns out that this makes little difference as long as the simulations with the neutral model use the same number of events as are recognized in the real data. If the four nonsignificant events are in fact spurious artifacts of sampling, they will only have the effect of adding random noise to the record and thus make rejection of the neutral model

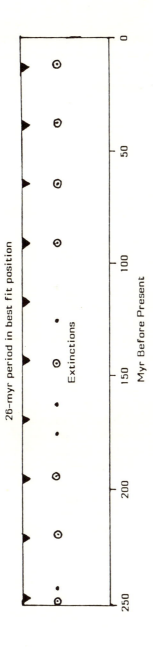

Figure 6.2. Estimated positions in time of 12 extinction events for marine families compared with a stationary periodic signal in best fit position. (Data from Raup and Sepkoski 1984, and Sepkoski and Raup 1986.)

less likely. It can be argued, therefore, that using all 12 events is the more conservative strategy.

Although the neutral model can be rejected, does this prove the alternative hypothesis of periodicity? The answer is that it does not! Rejection of the neutral model says only that the distribution is not random; it cannot say positively that the distribution is controlled by some other particular model because there is an uncountable number of alternative models that could be tested. On the other hand, the stationary periodicity is about the simplest alternative to a random distribution, and when given a choice, the data choose periodicity. It is on this basis that periodicity in extinction is claimed.

EQUILIBRIUM DIVERSITY IN ORDOVICIAN FAUNAS

Following the publication of MacArthur and Wilson's *Theory of Island Biogeography* (1967), a number of attempts were made to apply equilibrium theory to paleontological situations. One of the more ambitious was an analysis of benthic marine invertebrates of Ordovician age by Bretsky and Bretsky (1976). At hand were species lists for 215 horizons covering some 2,500 feet of shales and siltstones, equivalent to about five million years of deposition.

The Bretsky and Bretsky analysis was based on monitoring appearances and disappearances of species through the time sequence. A disappearance was defined as the absence of a species at a horizon when that species had been present at the next older horizon and an appearance was defined as the record of a species at a horizon when that species had been absent at the next older horizon. Disappearances and appearances were interpreted as extinctions and immigrations, respectively. For each assemblage, counts of total diversity (number of species), extinctions, and immigrations were recorded and analyzed.

Regression analysis showed that the number of extinctions increased linearly with diversity and the number of immigrations decreased linearly with diversity, and this is illustrated by the pair of solid lines in Figure 6.3. Bretsky and Bretsky concluded that the diversity value where the two regression lines intersect represents an equilibrium value in the style of MacArthur and Wilson. The analysis was performed separately for the five 500-foot intervals and all showed virtually the same pair of regression lines and the same equilibrium diversity. The authors concluded that faunas were maintained in dynamic equilibrium throughout the five-million-year interval even though there were rather marked changes in trophic composition.

This conclusion can be tested by applying a neutral model in which the MacArthur-Wilson equilibrium is explicitly absent. One way to do this is to scramble the time sequence of actual fossil assemblages and then analyze them with the Bretsky and Bretsky procedure. Bretsky and Bretsky (1976, Fig. 3) published all the species data for the 42 horizons sampled in the lowest 500 feet of section. The reordering of the data was a simple matter of randomizing the time sequence. This procedure retains the scale and fabric of the original data yet makes nonsense of the sequence of extinctions and immigrations. The neutral model is thus fo-

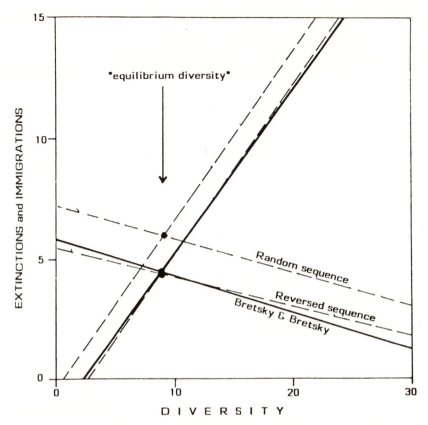

Figure 6.3. Equilibrium interpretation of fossil faunas. Solid lines: regressions of extinction versus diversity and immigration versus diversity for Ordovician assemblages. (From Bretsky and Bretsky, 1976, Fig. 6.) Intersection of the two lines is thought to represent an equilibrium diversity. Dashed lines: regressions based on randomized and reversed versions of the same data.

cused to neutralize the MacArthur-Wilson dynamics while leaving all other elements intact.

The exercise was performed also by a yet simpler neutral model. The temporal sequence of the real data for the 42 horizons was reversed so that time was forced to move backward. An immigration was recorded when a species occurred in a given sample but not in the sample that was originally above it in the sequence.

The results of the analyses of the randomized data were identical, within sampling error, to those obtained from the real data. This is indicated by the two pairs of dashed lines in Figure 6.3. Both kinds of simulation yield virtually the same "equilibrium diversity," and the regression lines for the reversed time sequence are essentially colinear with the Bretsky lines. The regression lines for the fully randomized time sequence are above the others because a modest amount of temporal memory present in the original data was removed by the randomization process.

The results just described can be shown to be true analytically by deriving the following equation for the general case:

$$\hat{S} = \frac{\Sigma S}{n} + \frac{\Sigma I - \Sigma E}{n(a_E - a_I)}$$

where $\hat{S}$ is equilibrium diversity (the intersection of the two regression lines), n is the number of horizons sampled, S is the observed diversity for a horizon, I is the number of immigrations at a horizon, E is the number of extinctions at a horizon, and a_E and a_I are the slopes of the two regression lines.

Because the total numbers of extinctions and immigrations are about the same, the numerator of the second term of the right side of the equation is approximately zero, making this term trivial. This means that the calculated value of equilibrium diversity approximates the sum of all diversities divided by the number of samples; that is, the mean diversity for the entire sequence. In fact, the equilibrium diversity found by Bretsky and Bretsky is numerically identical to actual mean diversity. The appearance of a MacArthur-Wilson equilibrium is thus an inevitable outcome of the structure of the original analysis. Further commentary on the Bretsky and Bretsky analysis has been provided by Lockley (1978) and Raab (1980), and these authors come to essentially the same conclusions with regard to the artifactual nature of the equilibrium diversity.

This case illustrates how a simple neutral model can be used to test a proposed biological explanation for a pattern seen in paleobiological data. In the particular case described here, the neutral model voids the biological explanation that had been proposed. If the results had turned out otherwise, the biological explanation would not have been proved but its basic credibility would have been supported.

EVOLUTIONARY FILLING OF MORPHOSPACE

Organic evolution over several billion years has produced a vast array of phenotypes, yet this diversity is generally thought to be less than could have developed. That is, the number of biologically possible phenotypes is much larger than the number actually seen in evolution. It is occasionally possible to glimpse this discrepancy, given a coordinate system in which all possible types can be defined and visualized. Figure 6.4 shows an example drawn from the coiled molluscs. Here, a two-dimensional morphospace is used to define the range in shape for nonheteromorph shelled cephalopod molluscs, and the distribution of 400 extinct genera is mapped in this space.

Figure 6.4 includes extinct cephalopods of the order Ammonoidea. For each taxon in the sample of 400 genera, the two critical measurements were made on one specimen, thus ignoring intraspecific and intrageneric variation. The contoured surface describes varying density of the 400 points in morphospace. As can be seen, most occurrences are concentrated in a relatively small region, leaving large areas empty, or virtually so. If subsets of the data, such as families or superfamilies, are plotted separately, the pattern becomes patchy. Indeed, patchy distributions are typical of such maps.

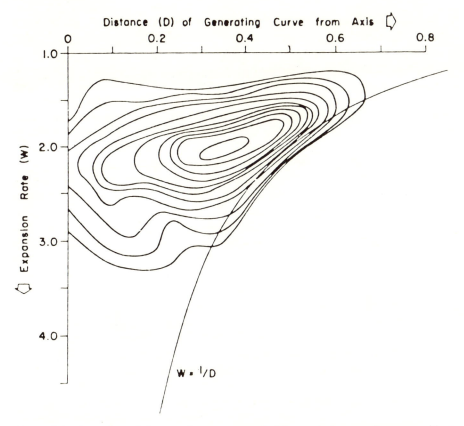

Figure 6.4. A portion of the morphospace occupied by nonheteromorph ammonoid cephalopods with a contoured distribution of occurrences of 400 ammonoid genera. (Modified from Raup 1967.)

How should the nonuniform filling of morphospace be interpreted? There are several deterministic interpretations, including: (1) insufficient geologic time available for complete and uniform filling, (2) concentration in certain parts of morphospace because these parts represent adaptive optima, and (3) presence of areas of morphospace which are "forbidden" because of biomechanical, developmental, or other constraints. On the other hand, the clustering may stem from an evolutionary process which is fundamentally aimless (random).

How do we make sensible and rigorous choices among these alternatives? In general, the second deterministic alternative (adaptive optima) is chosen, although this is commonly done in conjunction with the third alternative (forbidden areas). In the ammonoid case, the original interpretation (Raup 1967) was that the curved line that passes through morphospace in Figure 6.4 is a strong barrier because of biomechanical, functional, and other problems experienced by swimming cephalopods with shapes below this line. Given this constraint, the concentration of forms in part of the region above the line was interpreted as approximating the optimal geometry for these organisms. The case, while strong, is somewhat circumstantial and benefits from one's general faith in a strong adaptational paradigm: The conviction that the sheer presence of so many organisms of one shape demands an adaptational interpretation. This may well be correct but its testability is less than ideal.

One way to tackle the problem of uneven filling of morphospace is to devise a neutral model and ask, What would morphospace look like if none of the postulated biological mechanisms were operating? The neutral model is thus completely abiological in that there is no adaptation and there are no developmental, biomechanical, or other constraints: Evolving systems are free to move through morphospace at will. Only if the neutral model is unable to reproduce patterns actually seen in nature are the proposed biological mechanisms credible.

This approach appears at first glance to be feasible. A few characteristics of the neutral model are obvious. For example, the model should not be one where "species" are simply dropped at random on the surface of the morphospace, because to do so would destroy a fundamental feature of evolution whereby cladogenesis and phyletic transformation constrain a given species to be reasonably close to its immediate ancestor ("descent with modification"). Having coped with problems such as this, one could devise a Monte Carlo procedure to evolve imaginary ammonoid cephalopods in the format of Figure 6.4. The product would be a simulated filling of morphospace. A large number of such simulations would provide a statistical description of the kinds of pattern of morphospace filling that are likely.

The procedure just described has been attempted but never fully implemented because of one fundamental problem. In order to program the process of evolution necessary to fill morphospace, it is necessary to specify aspects of the process that are fundamentally biological but poorly understood. For example, are the evolutionary "steps" all the same size? Clearly they are not because to claim this would be to eliminate all variation in evolutionary rates (cladogenetic or phyletic). Therefore, step size might be drawn randomly from a hypothetical probability density distribution. If so, what would the shape of this distribution be? If the distribution were Gaussian, the result would be a two-dimensional random walk (Brownian motion) in morphospace. On the other hand, one could accommodate occasional large jumps by using a Cauchy instead of the Gaussian distribution. The result would be a much more patchy filling of the morphospace. This would probably look more like real patterns but would not prove anything about the real world unless one could justify the Cauchy distribution of step sizes as the inevitable and only way organisms evolve.

The difficulty just described demonstrates that the results of a neutral model of morphospace filling would be equivocal. That is, the results would be too model-dependent for them to tell us anything of interest. In other words, a truly neutral model cannot be constructed in the present state of knowledge of the evolutionary process. We are unable to neutralize the biological factors we are trying to study.

This case illustrates a situation where the strategy of building neutral models in paleobiology has failed. But it may be constructive in the long run because it focuses attention on shortcomings of the theoretical framework in which the filling of morphospace is investigated.

CONCLUSIONS

The four cases discussed illustrate a variety of outcomes of analysis using neutral models. In the trilobite case, the neutral model of random extinction was rejected, and this made credible the search for real biological causes of the extinction. In the case of periodic extinction, the neutral model was also rejected, and this added to the evidence for a nonrandom distribution of extinction events in geologic time. By contrast, the neutral model applied to Ordovician faunas showed that the deterministic explanation that had been proposed was not viable because any combination of the data would produce the same result. Finally, the approach failed in the case of evolutionary filling of morphospace because of the lack of a reasonable and workable neutral model.

These examples demonstrate that neutral models sometimes work positively and sometimes negatively with respect to the biological or other mechanisms that have been proposed to explain natural patterns. And in some cases, the approach cannot be implemented at all because the appropriate neutral model is inaccessible.

REFERENCES

Bretsky, P.W., and S.S. Bretsky. 1976. The maintenance of evolutionary equilibrium in Late Ordovician benthic marine invertebrate faunas. *Lethaia* 9:223–233.

Fischer, A.G., and M.A. Arthur. 1977. Secular variations in the pelagic realm. In *Deep Water Carbonate Environments,* ed. H.E. Cook and P. Enos, pp. 19–50. Society of Economic Paleontology and Mineralogy, Special Publication 25.

Lockley, M.G. 1978. The application of ecological theory to paleoecological studies, with special reference to equilibrium theory and the Ordovician System. *Lethaia* 11:281–291.

MacArthur, R.H., and E.O. Wilson. 1967. *The Theory of Island Biogeography.* Princeton: Princeton University Press.

Raab, P.V. 1980. Equilibrium theory and paleoecology. *Lethaia* 13:175–181.

Rampino, M.R., and R.B. Stothers. 1984. Terrestrial mass extinctions and galactic plane crossings. *Nature* 308:709–712.

Raup, D.M. 1967. Geometric analysis of shell coiling. *Journal of Paleontology* 41:43–65.

Raup, D.M. 1976. Species diversity during the Phanerozoic: A tabulation. *Paleobiology* 2:279–288.

Raup, D.M. 1981. Extinction: Bad genes or bad luck? *Acta Geologica Hispanica* 16:25–33.

Raup, D.M., S.J. Gould, T.J.M. Schopf, and D.S. Simberloff. 1973. Stochastic models of phylogeny and the evolution of diversity. *Journal of Geology* 81:525–542.

Raup, D.M., and J.J. Sepkoski, Jr. 1984. Periodicity of extinctions in the geologic past. *Proceedings of the National Academy of Sciences, USA* 81:801–805.

Raup, D.M., and J.J. Sepkoski, Jr. 1986. Periodic extinction of families and genera. *Science* 231:833–836.

Sepkoski, J.J., Jr. 1982. A compendium of fossil marine families. *Milwaukee Public Museum Contributions* 51:1–125.

Sepkoski, J.J., Jr., and D.M. Raup. 1986. Periodicity in marine extinction events. In *Dynamics of Extinction,* ed. D. Elliott, New York: John Wiley & Sons, pp. 3–36.

Stanley, S.M., P.W. Signor III, S. Lidgard, and A.F. Karr. 1981. Natural clades differ from "random" clades: Simulations and analyses. *Paleobiology* 7:115–127.

Valentine, J.W., T.C. Foin, and D. Peart. 1978. A provincial model of Phanerozoic marine diversity. *Paleobiology* 4:55–66.

7
Neutral Model of Taxonomic Diversification in the Phanerozoic: A Methodological Discussion

ANTONI HOFFMAN

One of the major research programs in modern paleobiology involves analysis of the pattern of taxonomic diversification through geological time. The best example of this approach is the analysis of the pattern of changes in marine animal family diversity in the Phanerozoic. The most commonly accepted causal explanation for this evolutionary pattern refers to a logistic model derived explicitly from the theory of island biogeography and extrapolated to the family level of taxonomic hierarchy and to the evolutionary time scale. My purpose in this article is to evaluate a rival explanation for this pattern, namely, a simple stochastic model that is neutral in the sense of accepting as few biological assumptions as possible. I argue that given the inherent limitations of the paleontological data, this neutral model cannot be rejected by the available empirical evidence. This conclusion leaves us on the horns of a methodological dilemma: How do we choose between two competing explanations for a unique historical pattern?

MACROEVOLUTIONARY MODEL OF DIVERSIFICATION

The pattern of diversity of marine animal families in the Phanerozoic is plotted in Figure 7.1. This diagram is based entirely on the data compiled by Sepkoski (1982) but it differs in some minor aspects from the curve presented recently by Sepkoski (1984, Fig. 1).

First, the number of families is here plotted against geological, instead of absolute, time; that is, the abscissa is calibrated in stratigraphic stages (or their equivalents, as used by Sepkoski 1982) rather than in millions of years. The original Sepkoski data have, in fact, been compiled at the stage level of stratigraphic resolution, and my way of plotting avoids introduction of the errors in absolute dating of stratigraphic boundaries to the data base. A change of the time scale inevitably results in considerable changes in the slope of the curve, especially in the Paleozoic, but it does not otherwise affect the pattern.

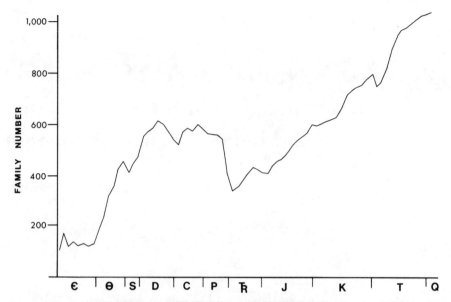

Figure 7.1. Number of fossil marine animal families per stratigraphic stage (or equivalent) in the Phanerozoic; calculated as the number of families surviving from the preceding stage plus half the families with the first record in the given stage minus half the families with the last record in the stage; time scale expressed in stratigraphic stages, not in absolute time units. (Data from Sepkoski 1982.)

Second, my diagram represents the entire data set covered by Sepkoski's compendium, including also poorly skeletonized taxa as well as the archaeocyathids and other problematic taxa, which have been omitted by Sepkoski (1984). The rationale for this decision of mine is that there is no unequivocal criterion to draw the distinction between well and poorly fossilizable organisms, especially since this distinction depends not only on the properties of the skeletal parts but also on the habitat of the organisms. The bias of differential fossilization potential equally affects well-understood fossil organisms and problematic taxa. On the other hand, the stratigraphic range of a taxon is always tentative, because it can be extended in either direction by finding its new fossil representatives, and it can also be expanded or contracted by new taxonomic interpretations. The likelihood of being subject to substantial taxonomic revision is essentially the same for poorly and well-skeletonized taxa. The inclusion of the entire data set, rather than its culled version, in the analysis introduces some changes to the absolute values of diversity, especially in the Cambrian and later Paleozoic, but the general shape of the curve remains very much the same. The number of families in each stratigraphic stage, however, is estimated by extending the family ranges between their first and last documented appearances (no matter if there are any actual records in between). Therefore, a consideration of the entire data set tends also to smooth out the curve of family diversity through geological time.

In order to account for the empirical pattern of global diversity of marine animals in the Phanerozoic, Sepkoski (1978, 1979) proposed a simple kinetic model developed by direct extrapolation of Rosenzweig's (1975) concept of continental steady-state diversity. Sepkoski suggested that the probabilistic (or per taxon) rate of origination of new high-rank taxa (orders or families) declines, and the probabilistic rate of extinction of taxa rises, with increasing diversity of the biota. This converse relationship of taxonomic evolutionary rates to diversity should lead to an evolutionary equilibrium of the number of taxa. When taken in conjunction with the concept of three evolutionary faunas—the Cambrian, the Paleozoic, and the modern one (Sepkoski 1981)—this model can be modified to represent three distinct but interacting groups of taxa. Each group, or evolutionary fauna, diversifies logistically according to its own set of parameters, but when the total diversity of this heterogeneous system exceeds the equilibrium diversity of any particular evolutionary fauna, the interference term in the model causes that fauna to decline and be gradually replaced by another fauna (Sepkoski and Sheehan 1983). The model can be further refined by introduction of extrinsic perturbations, modeled after the actual mass extinctions, which affect differentially the particular evolutionary faunas; and it then offers a very good fit to the empirical pattern of marine animal family diversity in the Phanerozoic (Sepkoski 1984). A further level of sophistication has been introduced by Kitchell and Carr (1985). They considered this system of three evolutionary faunas in terms of the logistic difference, instead of differential, equation and also in conjunction with major evolutionary innovations that reset the ecological system of interactions and thus establish a new equilibrium diversity.

Thus, the empirical pattern of taxonomic diversification can be successfully accounted for by a theoretical model. The model is macroevolutionary. I use the term macroevolutionary not only in the sense that the model is intended as explanation for a pattern observed at a high taxonomic level and on the evolutionary time scale; but also in that the model envisages processes that actually operate at that level of taxonomic hierarchy and at that time scale, that is, beyond the scope of natural selection and interspecific ecological interactions. In other words, acceptance of this model as the explanation for the pattern of Phanerozoic diversification would imply that the synthetic theory of evolution is incomplete and in need of substantial expansion.

The latter conclusion is conceptually plausible, but before it is accepted, the validity of the macroevolutionary model of diversification as *the* explanation for the empirical pattern should be firmly established. The trouble with this model, however, is that its parameters are not, and cannot be, independently estimated, because the empirical pattern is unique; that its fundamental premise, the assumption of diversity dependence of the rates of origination and extinction of taxa, is questionable (Walker and Valentine 1984; Hoffman 1985a). Therefore, matching the model to the empirical pattern is an exercise in curve fitting rather than in model testing (Hoffman 1981, 1983, 1985c). Evaluation of rival models thus becomes crucial.

NEUTRAL MODEL OF DIVERSIFICATION

The empirical pattern of taxonomic diversity in the Phanerozoic is the net result of two counteracting processes: the origination and the extinction of taxa. The patterns of marine animal family origination and extinction, derived from Sepkoski's (1982) data, are plotted in Figure 7.2. I propose that these patterns of origination and extinction can be adequately described as two independent random walks. In other words, I propose a neutral model of taxonomic diversification which assumes that the probabilistic rates of origination and extinction of taxa vary at random and independently of each other, but with the mean rate of origination exceeding the mean rate of extinction (Hoffman 1986).

Raup and Crick (1981) discussed the strategy of evaluation of a random walk hypothesis for a time series. Their testing procedure focuses on the number and length of runs, the number of positive and negative increments, and the mean value and frequency distribution of increments where increment is defined as the difference between two consecutive points in a time series, and run as the series of steps between two consecutive reversals in increment direction. These characteristics of the patterns of origination and extinction of taxa in the Phanerozoic should be analyzed in order to test the neutral model of diversification. Independence of these two patterns should also be analyzed.

Prior to evaluation of the neutral model, however, the empirical and theoretical limitations of the resolution potential of any analysis of taxonomic diversification in the Phanerozoic have to be briefly discussed because they severely constrain the testing procedure. As pointed out by Raup (1972, 1976b) and others (Sepkoski 1976; Lasker 1978; Signor 1978, 1982), compilations of global taxonomic diversity in the fossil record are very strongly biased by a variety of factors. These include the pull of the Recent (a Neogene fossil is more likely than a Cambrian one to be assigned to a Recent taxon) and the variation among geological time intervals in their absolute duration, outcrop area and/or sediment volume, fossiliferousness, number of *Fossil-Lagerstätten*, fossilization and recovery potential of the dominant fauna, and monographic coverage. Sepkoski's (1982) compen-

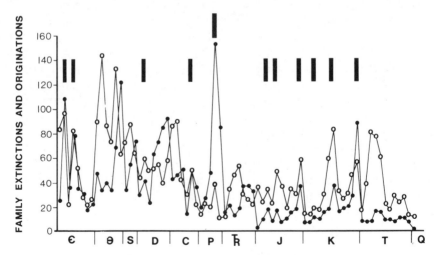

Figure 7.2. Patterns of the number of fossil marine animal family originations (open circles) and extinctions (closed circles) per stratigraphic stage (or equivalent) in the Phanerozoic; stages with coinciding peaks of family origination and extinction indicated. (Data from Sepkoski 1982.)

dium is further biased by the use of an analog of the ecological minimum census technique for determination of stratigraphic ranges and, consequently, for diversity estimation. This technique underestimates diversity near the beginning and the end of a time series of observations (Rosenzweig and Taylor 1980). By implication, Sepkoski's compendium and the consequent empirical pattern of Phanerozoic diversity (Fig. 7.1) are biased against high diversity in the Cambrian and Early Ordovician; in the geological context, the Recent data may be considered as an adequate measure of actual diversity, and therefore the Cenozoic data are not affected by the use of a minimum census technique. The reality of the most general features of the empirical pattern of diversity is supported by significant intercorrelation of several estimates of Phanerozoic diversity (Sepkoski et al. 1981). However, the majority of these estimates, those given by Raup (1976a, 1978), Seilacher (1977), and Sepkoski (1981), are all affected by the above biases. The only other estimate, the one given by Bambach (1977), is based on the average species number within individual communities. The choice of communities, however, is biased against high-diversity communities of the later Paleozoic (Hoffman 1985c); and a wide variety of small shelly fossils constituting a large, perhaps even dominant in terms of taxonomic diversity, component of the Cambrian biota (Rozanov 1986) are excluded from the analysis. All these biases imply that the data set on Phanerozoic diversity must be viewed and analyzed with caution because much of the variation in taxonomic diversity within, and probably also between, geological systems may represent merely the sampling error.

Further, and perhaps even more severe, constraints on testing of the models of taxonomic diversification in the Phanerozoic are imposed by our limited understanding of the processes of origination and extinction of taxa. It is unknown to what extent, and at what levels of time resolution, these processes operate continually or episodically. This uncertainty affects the choice of appropriate metrics of origination and extinction. If taxa originate and become extinct continually, at least at the time scale of a million years, the patterns of family origination and extinction presented in Figure 7.2 should be normalized for time in order to obtain the rates of origination and extinction per million years, because, clearly, longer time intervals would, then, accumulate more events of family origination and extinction. This is, in fact, the assumption underlying the recent analyses of secular change in family extinction rates in the Phanerozoic (Raup and Sepkoski 1982; Van Valen 1984). The difficulty with such time-normalized metrics of origination and extinction is that the margins of uncertainty about the actual duration of particular stratigraphic stages are so wide (Odin 1982; Harland et al. 1983; Snelling 1985) that the rates per million years may vary by an order of magnitude depending on the time scale adopted, and the identity and timing of their peaks may undergo considerable shifts (Hoffman 1985b).

To overcome the problem with absolute dating of stratigraphic boundaries, the rates of extinction and origination per stratigraphic stage might be accepted as the appropriate metrics. This would assume that the processes of origination and extinction of taxa are episodic. Such is, in fact, the premise underlying the recent analysis of periodicity in mass extinctions (Raup and Sepkoski 1984; see also Raup, Chapter 6). However, even if the apparent periodicity of mass extinctions is assumed to indicate an extraterrestrial mechanism of episodic extinction (but see Kitchell and Pena 1984; Hoffman and Ghiold 1985; Hoffman 1985b), such a conclusion applies to only a fraction of points in the time series to be evaluated. Consequently, its acceptance would necessitate time resolution of the data on global rates of extinction at a much finer stratigraphic scale than that employed by Sepkoski (1982). At the moment, this is far beyond the resolution potential of stratigraphic correlation for most of the Phanerozoic.

This discussion implies that the empirical patterns shown in Figure 7.2, and consisting of the numbers of events per stratigraphic stage, represent a distorted picture of the actual patterns of family origination and extinction through time. The extent and the direction of distortion are unknown. Therefore, testing of these patterns for the mean value and the frequency distribution of increments in the time series is impossible (see discussion on statistical identification of mass extinctions as outliers in the time series of extinction rates in the Phanerozoic; Raup and Sepkoski 1982; Quinn 1983; Raup et al. 1983; Stigler, Chapter 8). This is even more so because of the decline in rates of family extinction (Raup and Sepkoski 1982; Van Valen 1984; Kitchell and Pena 1984) and origination (Hoffman 1986) in the Phanerozoic. The slope of this decline depends very strongly on the choice of the metric and the time scale, which consequently affect also analyses of the residuals and their frequency distribution.

The decline in rates of origination and extinction in the Phanerozoic might be taken as evidence against the neutral model of diversification because it seems to be incompatible with the random walk hypothesis of origination and extinction of taxa. It appears to suggest some biological controls on these processes at the global scale. Raup and Sepkoski (1982) suggested, in fact, that this decrease in rate of extinction implies an increase in overall fitness (meant to denote adaptedness rather than genetic fitness) of the biota, whereas Van Valen (1984) interpreted it as an indication of decreasing ecological interference and increasing cooperation among taxa over geological time.

However, the decline occurs at the family level of taxonomic hierarchy and may be compatible with stochastic constancy of the rates of origination and extinction at the species level. There is no necessary congruence between evolutionary patterns at different levels of taxonomic hierarchy. In fact, some incongruence is to be expected (Holman 1983). Three major factors inevitably affect this relationship and may contribute to the secular decline of family rates of origination and extinction even under stochastic constancy of the species rates. First, the hierarchic taxonomic procedure imposed on the branching evolutionary tree results in higher-rank taxa reaching their maximum diversity earlier than lower-rank taxa (Raup 1983). Thus, family origination rates may decline through time even when speciation rates are constant, simply reflecting an artifact of the taxonomic structure of the biosphere. Second, the average marine animal family tends to be more and more speciose through time; this phenomenon may account for decline in family extinction rates in spite of a stochastic constancy in species extinction rates (Flessa and Jablonski 1985). Finally, the pull of the Recent (Raup 1972) tends to artificially suppress more recent extinction events and to extend the actual stratigraphic ranges of taxa backwards, thus decreasing both the family origination and extinction rates in progressively younger intervals of geological time.

Under unknown quantitative contributions of these three biasing factors, the observed decline in family origination and extinction rates may be biologically meaningless. Consequently, however, any analysis of the absolute values of the rates of origination and extinction of taxa must remain inconclusive.

Thus, testing of the neutral model of taxonomic diversification in the Phanerozoic must be focused on the numbers of positive and negative increments and the number and length of runs in, and the mutual relationship of, the empirical patterns of origination and extinction rates shown in Figure 7.2.

These patterns pass the test for random walk (Hoffman and Ghiold 1985; Hoffman 1986). Insofar as their interrelationship is concerned, a correlation analysis between time series could be misleading; this is especially true for the global patterns of origination and extinction because many biases of the fossil record affect simultaneously and similarly both the patterns. Independence of two random walks can be tested by analysis of the location of their peaks, because the probabilities of particular configurations (peak in one random walk occurring one step before, at the same step as, or one step after a peak in the other random walk) are equal; in contrast, one configuration would be privileged if the time series were somehow intercorrelated. The patterns of family origination and extinction in the Phanerozoic significantly deviate from this prediction of the neutral model only with respect to the number of coincident peaks. However, this deviation appears to be due to the bias of *Fossil-Lagerstätten*. An unusual accumulation of otherwise nonfossilizable organisms allows for the recognition of a large number of taxa known exclusively from a single site, although their actual stratigraphic ranges must have extended beyond the deposition of these particular strata. Consequently, both the rates of origination and extinction become artificially inflated. The geological stages with coincident peaks of family origination and extinction (Figure 7.2) include all the most famous marine *Fossil-Lagerstätten,* that is, the Middle Cambrian Burgess Shale, Lower Devonian Hunsrück Slate, Carboniferous Mazon Creek, and Upper Jurassic Solnhofen Limestone. When the taxa known exclusively from these *Fossil-Lagerstätten* are deleted from the data set, the Middle Cambrian and the Carboniferous coincidences of peaks of origination and extinction disappear, and the patterns become entirely consistent with the model of two independent random walks. This is the case with the strict random walk model, with .5 probability of going up and down at each step in the time series (Hoffman and Ghiold 1985; Hoffman 1986). If a model is chosen with points drawn at random from either normal or Poisson distribution, the expected number of peaks in a time series is slightly greater. The two empirical patterns conform then at P = .05 to the model.

I conclude that given all the limitations and biases of the fossil record, the neutral model of diversification withstands the test of empirical data.

DISCUSSION

As a null hypothesis, the neutral model of taxonomic diversification cannot be rejected by the available evidence. This result may be due to the insufficient understanding of the processes of origination and extinction of taxa, which makes impossible the choice of appropriate metric, the precise definition of appropriate neutral model, and its adequate, rigorous testing. This result may also be due to the inherent limitations of the data set, which is so severely affected by various sampling biases, stratigraphic uncertainty, and arbitrary taxonomic decisions that the statistical noise may overwhelm the actual orderly pattern, if there is one. In other words, the neutral model may be too null to be rejected, at least at the present state of theoretical and empirical knowledge about origination and extinction of taxa. But the conclusion that the empirical pattern of taxonomic diversification can be described by the neutral model may also indicate true randomness of the actual pattern of diversity in the Phanerozoic. Thus, the present analysis does not demonstrate that the neutral model is correct, but only that there is no reason to reject it. It provides a viable explanation for the empirical pattern.

Even if the neutral model is correct and the analyzed pattern of taxonomic diversification is indeed explained and not merely described by randomness and mutual independence of the underlying biological processes of origination and extinction of taxa, the model accounts only for the general pattern. That is, the presumed correctness of the neutral model would not imply that individual events (e.g., mass extinctions or waves of origination) jointly constituting the pattern represent pure chance, because randomness may also arise at the intersection of a very large number of deterministic but essentially independent processes. The Late Permian and terminal Cretaceous mass extinctions would still demand a causal explanation, as would also the Cambro-Ordovician and Paleogene explosive radiations. But the entire pattern of diversity in the Phanerozoic would have to be considered as the net result of a random series of events, like the series of heads and tails while flipping a coin.

The caveat about inconclusiveness of this analysis of the neutral model as explanation for the pattern of taxonomic diversification is not only a routine methodological qualification, intended to avoid the type II statistical error (acceptance of false hypothesis) and, therefore, always relevant to hypothesis testing. Given all the notorious limitations of the fossil record, the macroevolutionary model of taxonomic diversification also withstands the test as a null hypothesis. It can be fitted to the empirical data (Sepkoski 1984) and the most serious reservation concerning this model implies only the absence of any empirical support for, but not the ultimate refutation of, its assumption of diversity-dependence of the rates of origination and extinction (Hoffman 1985a).

We thus face a real methodological dilemma. When regarded as null hypotheses, both models withstand the test. This may be due to the sloppiness of the available data, but there is little hope that we will be able to collect adequate data soon. How to decide which one of the two rival models should be accepted, at least provisionally, as the explanation for the empirical pattern? What methodological criteria to use in order to solve this problem?

I suggest that this dilemma exemplifies a more general methodological issue, the one concerning the status of the null hypothesis and its testing as the prime method in science.

The methodological reliance on the null hypothesis as the main tool of scientific analysis goes back to Popper's (1959) falsificationism. It rests on the belief that the logical asymmetry between verification and falsification of a hypothesis translates into their methodological asymmetry. In principle, ultimate verification is impossible because all consequences of a theory can never be identified, articulated, and put on test. Ultimate falsification, in turn, is logically possible because if only one consequence of a theory is false, the theory itself must be rejected. This logical asymmetry, however, may justify the methodological falsificationism if, and only if, there exist rules or criteria allowing for a clear and unequivocal demonstration that a consequence of the theory being tested is indeed false. In other words, the methodological asymmetry between verification and falsification is contingent upon the existence of an *experimentum crucis*.

The latter premise, however, appears to be invalid (Amsterdamski 1983). No hypothesis can ever be tested in isolation, separated entirely from the context of other theories and hypotheses. Therefore, a disagreement between hypothesis and empirical results does not falsify the hypothesis, because some other elements of the logical conjunction being compared to empirical data may be false. It is always possible that the hypothesis is correct but it nevertheless results in a false prediction because it is taken in conjunction with incorrect associated theory. Similarly, a fit between hypothesis and empirical results does not demonstrate its correctness, because an associated theory may be false, whereas conjunction of two false statements may lead to a true conclusion.

The well-known thesis of Duhem (1954) postulates in fact that if an experiment is in disagreement with theoretical prediction, such a result indicates that at least one of the hypotheses or theories involved is false, but it does not identify the one that should be modified. The most common counterargument is that when we test two rival hypotheses in the context of the same group of associated theories, an *experimentum crucis* is possible because the identity of the context allows for recognition of the incorrect hypothesis (Popper 1963). This argument, however, would be valid only if all the associated theories were known to be correct, which is, of course, unlikely in any particular problem situation.

The Duhem thesis has been strengthened by Quine (1980), who postulated not only that an *experimentum crucis* never identifies the false hypothesis, but also that its negative results can always be reconciled with the tested theory by introduction of one or more *ad hoc* hypotheses. This claim, the Duhem-Quine thesis, has not been demonstrated to be correct in general, but neither has it been disproved. As pointed out by Amsterdamski (1983), the least condition that must be met to disprove it is that criteria be given that unequivocally identify *ad hoc* hypotheses as a class. Such criteria, however, do not exist (Hempel 1966; Grünbaum 1976). Consequently, a theory can be defended with use of what certain experts regard as well-grounded auxiliary arguments, but what others regard as unacceptable *ad hoc* hypotheses.

Whether the Duhem-Quine thesis is indeed universally valid in science may be debatable. I feel, however, that the models of taxonomic diversification in the Phanerozoic provide a clear illustration of this dilemma. As indicated by Sepkoski (1979), the macroevolutionary model of diversification must be rejected if the archaeocyathids are interpreted as marine animals and are included in the analysis. The model is, therefore, defended by considering them plants and excluding them from the data set. In its later, multiphase version (Sepkoski and Sheehan 1983; Sepkoski 1984), the model could also be defended by erecting a fourth evolutionary fauna, embracing the archaeocyathids and perhaps some other Early to Middle Cambrian organisms. The rates of family origination and extinction do not seem to be diversity-dependent, as assumed by the macroevolutionary model (Hoffman 1985a). But the model can still be defended by assuming some complex nonlinear functions, although there is no evidence, empirical or theoretical, for any particular form of these functions.

In its turn, the neutral model of diversification predicts a significantly smaller number of coinciding peaks of family origination and extinction than the number found in the empirical pattern. The model can be defended, however, by attributing this misfit to the systematic bias introduced by *Fossil-Lagerstätten*. The secular trends in the rates of family origination and extinction can be explained by the inferred incongruence between patterns at the family and the species levels (Hoffman 1986). Given the notorious inadequacy of the fossil record and the uniqueness of the empirical pattern of diversification, which rules out independent estimation of the models' parameters and allows for reference to unique boundary conditions, both the models probably can be defended endlessly.

However, there is no *a priori* logical reason why the testing of null hypotheses should be relied upon as the prime method of theory evaluation. Within the conceptual framework of falsificationism, even in a sophisticated falsificationism of Lakatos (1970), the null hypothesis is being put in a privileged position because it is at least tentatively accepted if it withstands the test (Van Valen 1982). That it should not be so follows inevitably from the Duhem-Quine thesis, and in fact even from its weaker version, the Duhem thesis, alone. This corollary is clearly exemplified by the problem with the pattern of taxonomic diversification, because neither of the two contrasting null hypotheses can be rejected by the available data.

Some other approaches and methodological criteria must be used to determine which one of the rival models should be provisionally accepted. One such approach refers to the concept of likelihood (Hacking 1965; Edwards 1972). The likelihood of a hypothesis indicates how well it is supported by a given set of empirical observations. A statistical comparison of the fit of competing hypotheses to the empirical data thus may indicate the best-supported hypothesis. It can, then, be decreed as a methodological rule that the hypothesis with the highest likelihood relative to all the available data should be accepted until another hypothesis— with still more likelihood—is proposed. For example, if a random sample of 10 balls is drawn from an urn and they all happen to be black, the hypothesis that 90% of all the balls in the urn are black offers a better fit to the data than the hypothesis that only 30% of the balls in the urn are black; but the hypothesis that all the balls in the urn are black gives the best fit. On the basis of the likelihood criterion, then, the last hypothesis should be accepted.

The likelihood criterion is debatable in principle because it consistently prefers *ad hoc* hypotheses based on curve fitting to empirical observations rather than general, and therefore somewhat less precise, explanations. Even if we knew that, as a rule, the company producing such urns did not sell urns with all balls identical, we still would be forced to accept the all-black hypothesis as the maximum likelihood one, and perhaps to support it with an *ad hoc* hypothesis about a human mistake or a deliberate joke. The likelihood criterion also gets into considerable practical trouble if the observations incorporate an unknown error component, because the comparison between hypotheses may, then, consist of measuring their fit to statistical noise. If the color of the balls drawn from the urn was determined by a color-blind person but we were so suspicious of the urn manufacturer that we preferred to rely more on this determination than on the company warranty, the likelihood criterion would lead us astray if the colors were indeed misidentified. Of course, the larger the error component of empirical data, the more acute is this problem; and this is precisely the case with the macroevolutionary and the neutral models of taxonomic diversification.

Another approach to the problem in choice between competing hypotheses refers to the concept of parsimony or, essentially, simplicity (Sober 1975). In principle, there is no reason to think that the simplest hypothesis is necessarily true. It is merely a matter of convention that either makes the parsimony an inherent part of rationality, or accepts it as a basic element of the scientific method in spite of its irrationality. But insofar as the criterion of parsimony can be formalized to provide an unequivocal solution in any particular problem situation, it can be adopted as a methodological rule.

It is, in fact, the concept of simplicity that constitutes the ultimate basis for the common preference for neutral models, which assume randomness as the explanation for empirical observations, rather than other causal explanations (e.g., Raup et al. 1973; Strong et al. 1979). For the randomness *per se* does not have any inherent logical priority relative to deterministic causality. It only appears to represent a simpler, more parsimonious explanation. Such a preference for simplicity, however, is valid only if the randomness is interpreted ontologically, as in the subatomic world where randomness is regarded as the sole causal agent. This is clearly not so in evolutionary biology, where randomness can only be interpreted operationally (Schopf 1979; Hoffman 1981) as the net result of multifaceted, independent processes, often additionally obscured by enormous observational errors.

Consequently, the criterion of simplicity alone cannot determine the preferability of either the macroevolutionary or the neutral model of taxonomic diversification. It has to be coupled with the criterion of hypothesis compatibility with the synthetic paradigm of modern evolutionary biology (Hoffman 1979). This paradigm is constituted by the concept of natural selection acting upon genetic mutations that arise at random relative to environmental challenges and the resultant adaptive needs of the organisms. This paradigm can be justified deductively (Van Valen 1982). It does not claim that all the evolutionary phenomena are explained by this process, but only that such a process *must* occur in nature. A model or theory aiming at explanation of a particular evolutionary pattern may, therefore, go beyond the synthetic paradigm, but in agreement with the criterion of simplicity, this is permissible only if there is a demonstrated need to do so, that is, if the empirical observations cannot be satisfactorily accounted for by the synthetic paradigm itself. This is the principle of *pragmatic reductionism* (Hoffman 1983; Schopf 1984).

Under this methodological principle, the neutral model of taxonomic diversification is clearly preferable to the macroevolutionary model, because it is entirely compatible with the synthetic paradigm. Its assumptions imply only that there are so many individual species in the biosphere that respond to so many environmental challenges that the average rates of speciation and species extinction at any particular time are unpredictable, although in the long run, the mean rate of speciation exceeds the mean rate of species extinction. The macroevolutionary model, in contrast, postulates the emergence of high-rank taxa and their groups (i.e., evolutionary faunas of Sepkoski 1981) as units of evolution, and the operation of evolutionary processes that are not reducible to natural selection and interspecific interactions. That the neutral model withstands the test thus demonstrates that the synthetic paradigm can account for the pattern of marine animal family origination and extinction in the Phanerozoic. There is no need for macroevolutionary laws in order to explain that pattern.

ACKNOWLEDGMENTS

I thank Gene Fenster and Max Hecht for discussion. This article was written under tenure of NSF grant BSR 84-13605.

REFERENCES

Amsterdamski, S. 1983. *Between History and Method* (in Polish). Panstwowy Instytut Wydawniczy, Warsaw.
Bambach, R.K. 1977. Species richness in marine benthic habitats through the Phanerozoic. *Paleobiology* 3:152–167.
Duhem, P. 1954. *The Aim and Structure of Physical Theory*. Princeton: Princeton University Press.
Edwards, A. 1972. *Likelihood*. Cambridge: Cambridge University Press.
Flessa, K.W., and D. Jablonski. 1985. Declining Phanerozoic background extinction rates: Effect of taxonomic structure? *Nature* 313:216–218.
Grünbaum, A. 1976. Ad hoc auxiliary hypotheses and falsificationism. *British Journal for the Philosophy of Science* 27:329–362.
Hacking, I. 1965. *The Logic of Statistical Inference*. Cambridge: Cambridge University Press.
Harland, W.B., A.V. Cox, P.G. Llewellyn, C.A.G. Pickton, A.G. Smith, and R. Walters. 1983. *A Geologic Time Scale*. Cambridge: Cambridge University Press.
Hempel, C.G. 1966. *The Philosophy of Natural Science*. Englewood Cliffs, N.J.: Prentice-Hall.
Hoffman, A. 1979. Community paleoecology as an epiphenomenal science. *Paleobiology* 5:357–379.
Hoffman, A. 1981. Stochastic versus deterministic approach to paleontology: The question of scaling or metaphysics? *Neues Jahrbuch für Geologie und Paläontologie, Abhandlungen* 162:80–96.
Hoffman, A. 1983. Paleobiology at the crossroads: A critique of some modern paleobiological research programs. In *Dimensions of Darwinism*, ed. M. Grene, pp. 241–271. Cambridge: Cambridge University Press.
Hoffman, A. 1985a. Biotic diversification in the Phanerozoic: Diversity independence. *Palaeontology* 28:387–391.

Hoffman, A. 1985b. Patterns of family extinction depend on definition and geological timescale. *Nature* 315:659–662.
Hoffman, A. 1985c. Island biogeography and palaeobiology: In search for evolutionary equilibria. *Biological Reviews* 60:455–471.
Hoffman, A. 1986. Neutral model of Phanerozoic diversification: Implications for macroevolution. *Neues Jahrbuch für Geologie und Paläontologie* 172:219–244.
Hoffman, A., and J. Ghiold. 1985. Randomness in the pattern of "mass extinctions" and "waves of origination." *Geological Magazine* 122:1–4.
Holman, E.W. 1983. Time scales and taxonomic survivorship. *Paleobiology* 9:20–25.
Kitchell, J.A., and T.R. Carr. 1985. Nonequilibrium model of diversification: Faunal turnover dynamics. In *Phanerozoic Diversity Patterns*, ed. J.W. Valentine, pp. 277–309. Princeton: Princeton University Press.
Kitchell, J.A., and D. Pena. 1984. Periodicity of extinctions in the geologic past: Deterministic versus stochastic explanations. *Science* 226:689–692.
Lakatos, I. 1970. Falsification and the methodology of scientific research programs. In *Criticism and Growth of Knowledge*, ed. I. Lakatos and A. Musgrave, pp. 91–196. Cambridge: Cambridge University Press.
Lasker, R.H. 1978. The measurement of taxonomic evolution: Preservational consequences. *Paleobiology* 4:135–149.
Odin, G.S., ed. 1982. *Numerical Dating in Stratigraphy*. New York: John Wiley & Sons.
Popper, K.R. 1959. *The Logic of Scientific Discovery*. London: Hutchinson.
Popper, K.R. 1963. *Conjectures and Refutations: The Growth of Scientific Knowledge*. New York: Harper & Row.
Quine, W.V.O. 1980. *From a Logical Point of View*. Cambridge, Mass.: Harvard University Press.
Quinn, J.F. 1983. Mass extinctions in the fossil record: Discussion. *Science* 219:1239–1240.
Raup, D.M. 1972. Taxonomic diversity during the Phanerozoic. *Science* 177:1065–1071.
Raup, D.M. 1976a. Species diversity in the Phanerozoic: A tabulation. *Paleobiology* 2:279–288.
Raup, D.M. 1976b. Species diversity in the Phanerozoic: An interpretation. *Paleobiology* 2:289–297.
Raup, D.M. 1978. Cohort analysis of generic survivorship. *Paleobiology* 4:1–15.
Raup, D.M. 1983. On the early origins of major biologic groups. *Paleobiology* 9:107–115.
Raup, D.M., and R.E. Crick. 1981. Evolution of single characters in the Jurassic ammonite *Kosmoceras*. *Paleobiology* 7:200–215.
Raup, D.M., S.J. Gould, T.J.M. Schopf, and D.S. Simberloff. 1973. Stochastic models of phylogeny and the evolution of diversity. *Journal of Geology* 81:525–542.
Raup, D.M., and J.J. Sepkoski. 1982. Mass extinctions in the marine fossil record. *Science* 215:1501–1503.
Raup, D.M., and J.J. Sepkoski. 1984. Periodicity of extinctions in the geologic past. *Proceedings of the National Academy of Sciences, USA* 81:801–805.
Raup, D.M., J.J. Sepkoski, and S.M. Stigler. 1983. Mass extinctions in the fossil records: Reply. *Science* 219:1240–1241.
Rosenzweig, M.L. 1975. On continental steady states in species diversity. In *Ecology and Evolution of Communities*, ed. M.L. Cody and J.M. Diamond, pp. 121–140. Cambridge, Mass.: Harvard University Press.
Rosenzweig, M.L., and J.A. Taylor. 1980. Speciation and diversity in Ordovician invertebrates: Filling niches quickly and carefully. *Oikos* 35:236–243.

Rozanov, A.Y. 1986. Problematica of the Early Cambrian. In *Problematic Fossil Taxa*, ed. A. Hoffman and M.H. Nitecki, pp. 87-96. New York: Oxford University Press.

Schopf, T.J.M. 1979. Evolving paleontological views on deterministic and stochastic approaches. *Paleobiology* 5:337-352.

Schopf, T.J.M. 1984. Rates of evolution and the notion of "living fossils." *Annual Reviews of Earth and Planetary Science* 12:245-292.

Seilacher, A. 1977. Evolution of trace fossil communities. In *Patterns of Evolution*, ed. A. Hallam, pp. 359-376. Amsterdam: Elsevier.

Sepkoski, J.J. 1976. Species diversity in the Phanerozoic: Species-area effects. *Paleobiology* 2:298-303.

Sepkoski, J.J. 1978. A kinetic model of Phanerozoic taxonomic diversity. I. Analysis of marine orders. *Paleobiology* 4:223-251.

Sepkoski, J.J. 1979. A kinetic model of Phanerozoic taxonomic diversity. II. Early Phanerozoic families and multiple equilibria. *Paleobiology* 5:222-251.

Sepkoski, J.J. 1981. A factor analytic description of the Phanerozoic marine fossil record. *Paleobiology* 7:36-53.

Sepkoski, J.J. 1982. A compendium of fossil marine families. *Milwaukee Public Museum Contributions in Biology and Geology* 51:1-125 (and corrections of August 1983).

Sepkoski, J.J. 1984. A kinetic model of Phanerozoic taxonomic diversity. III. Post-Paleozoic families and mass extinctions. *Paleobiology* 10:246-267.

Sepkoski, J.J., R.K. Bambach, D.M. Raup, and J.W. Valentine. 1981. Phanerozoic marine diversity and the fossil record. *Nature* 293:435-437.

Sepkoski, J.J., and P.M. Sheehan. 1983. Diversification, faunal change, and community replacement during the Ordovician radiations. In *Biotic Interactions in Recent and Fossil Benthic Communities*, ed. M.J.S. Tevesz and P.L. McCall, pp. 673-717. New York: Plenum Press.

Signor, P.W. 1978. Species richness in the Phanerozoic: An investigation of sampling effects. *Paleobiology* 4:394-406.

Signor, P.W. 1982. Species richness in the Phanerozoic: Compensating for sampling bias. *Geology* 10:625-628.

Snelling, N.J., ed. 1985. The chronology of the geological record. *Memoir of the Geological Society* (London) 10.

Sober, E. 1975. *Simplicity*. Oxford: Oxford University Press.

Strong, D.R., L.A. Szyska, and D.S. Simberloff. 1979. Tests of community-wide character displacement against null hypotheses. *Evolution* 33:897-913.

Van Valen, L.M. 1982. Why misunderstand the evolutionary half of biology? In: *Conceptual Issues in Ecology*, ed. E. Saarinen, pp. 323-344. Dordrecht: Reidel.

Van Valen, L.M. 1984. A resetting of Phanerozoic community evolution. *Nature* 307:50-52.

Walker, T.D., and J.W. Valentine. 1984. Equilibrium models of evolutionary species diversity and the number of empty niches. *American Naturalist* 124:887-899.

8

Testing Hypotheses or Fitting Models?
Another Look at Mass Extinctions

STEPHEN M. STIGLER

INTRODUCTION: A STATISTICIAN'S VIEWPOINT

A large number of the methodological terms that have been employed in empirical research in evolutionary biology are more than merely familiar to statisticians; we have a proprietary interest in them. The terminology of "testing null hypotheses" originated, I believe, in the statistical literature of the 1930s. Now I should hasten to add that no statistician would begrudge the use of such terms (and the associated ideas) by biologists, geophysical scientists, and philosophers of science—quite the contrary. They were in fact developed for export, and we would be more offended if you ignored our products than if you consumed them. Still, as with all products, there can come a time when experience or further research can lead to the need for updates, bulletins, or even, in drastic cases, product recall. Or the producer may learn of a new and unanticipated use that a product is being put to, and feel the need to issue a warning. And, of course, new products come along. We all know that old 78-rpm phonograph records still work as they were designed to work, but no one who has heard the new digitally encoded compact discs would want to return to 78s in any but an emergency situation. My purpose in this chapter is to comment on such issues, taking the role of a factory representative addressing an audience of consumers.

To begin with, a few comforting disclaimers: It will be beyond my intended scope to discuss new products, but my limited experience with applications of statistics in the fields under discussion suggests that biologists are not employing 78-rpm level techniques. They are at least up to the $33\frac{1}{3}$-rpm level, although occasionally their procedures are pressed out of rather cheap vinyl, and they might want, when time does permit, to visit a statistical shop and sample the new product line (all digital, of course!). And I have encountered few gross abuses in the literature; while they are not absent, they need not concern us here.

To these two disclaimers—that I shall not brag about new (possibly not adequately tested) products, and I shall not dwell upon a small minority of cases of product abuse—I should add a third: I will not issue a recall announcement. Statisticians' sins are no less than those of other disciplines, and we are no more eager to make public confessions and recantations about them than biologists may be, but as far as I know our product remains fundamentally sound. Instead, I bring a much less urgent factory bulletin, a list of a few principles that may be useful in maintaining a high level of performance with an aging, albeit solid, product.

Now my bulletin could, for example, focus upon the term "neutral" used so often in describing a null hypothesis (a term, I should hasten to say, that was attached to our product *after* it left the factory), and I could issue a string of general caveats. These might include:

1. Beware of false dichotomies; rejecting a straw null hypothesis does not prove a pet theory.
2. Beware of tests of low power and the fallacy of thinking that accepting a null hypothesis proves that it is true.
3. Beware of the problem of testing too many hypotheses; the more you torture the data, the more likely they are to confess, but confessions obtained under duress may not be admissible in the court of scientific opinion.
4. Beware of the siren lure of a supposedly nonparametric procedure that hides its Achilles heel in unfamiliar territory, be it in the Fourier domain or in a massive simulation.

I could give cautionary examples of what can happen if these caveats are ignored, some as current as last year's *Nature* or *Science* magazines, but there are two problems with this approach. First, negative messages are seldom heeded (model airplane glue tubes all warn the user not to inhale, and that warning serves the perverse role of suggesting that inhaling might be worth trying). And second, at that level of generality I would not be saying anything new. And so I have chosen to take a more positive approach, to illustrate a few more constructive principles, which, while they also may not be new, seem to be less well known than they should be.

MASS EXTINCTIONS VERSUS BACKGROUND EXTINCTIONS

I will be discussing these principles in the context of a real example, the original study of extinction rates by Raup and Sepkoski. Raup (Chapter 6) has discussed one aspect of this topic—the search for periodicities in extinction, a topic that has caught the public imagination in the last couple of years. Other literature on extinction is reviewed by Jablonski (1984). Although I will make a few comments upon the question of periodicity, my main focus is on a different methodological point, and I will consequently concentrate on a different aspect of the study, namely, on the degree to which the data permit a qualitative distinction between mass extinctions and background extinctions.

In 1982 Raup and Sepkoski published their first investigation of temporal patterns in extinction rates (Figure 8.1). Their purpose was twofold: to investigate the evidence for (and prevalence of) events of mass extinction in the marine fossil record, and to assess the apparent decline in rate of extinction. Now at the outset I should issue a warning of my own—in reconsidering these data my goal is

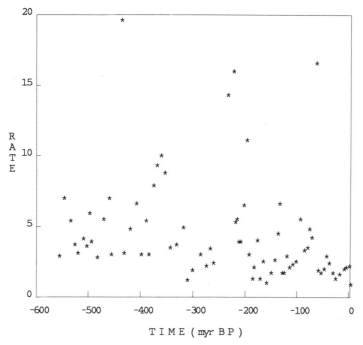

Figure 8.1. Extinction rate R (or number of extinctions per million years) for each of $N = 76$ stratigraphic stages, versus time in million years. (Based on Raup and Sepkoski 1982.)

primarily methodological, and I shall gloss over some points that should be attended to in a full analysis, although I believe the substance of my conclusions would hold up under a fuller scrutiny. I also am dealing with a version of their data that is three years old, and thus not taking advantage of significant advances in knowledge about both the time scale and the extinction dates, much of this due to Raup and Sepkoski themselves.

My aim in this discussion is to advance two constructive principles:

1. The choice of a "null model" (I resist the phrase "neutral model" for reasons that may be apparent later) depends crucially upon which question you ask of the data; the same data can and should lead to different null models when different questions are entertained.
2. More can be learned through an analysis that compares different plausible models than by simply testing one against the alternative, "everything else." A corollary to this is, "fit models, don't just test null hypotheses."

The data Raup and Sepkoski considered may be usefully, if simplistically, thought of as having been arrived at in the following manner: At discrete time points $t_1, t_2, \ldots$, individual families breathe their last (or whatever these marine animals do when they expire), but these times are known only (and at best) up to the level of the stratigraphic stage in which the most recent fossil has been found. Not all "last fossils" are datable to the level of the stage; I ignore that fact in this analysis, using the proportional allocation of lower resolution data that

Raup and Sepkoski used. I do not believe this will affect my conclusions, although it should be considered in a full analysis.

The first question to be considered is, What do, or can, these data tell us about mass extinctions? It is visually evident that some of these empirical rates are large, but are they unduly so? This is a question that cannot be answered except in relation to some hypothesis. Surely the large rates are large relative to a hypothesis that says the empirical rates are absolutely constant over time, that extinctions occurred at an absolutely uniform rate. But that would not be an interesting hypothesis, except perhaps to a group of eighteenth-century social philosophers who had rather restricted views as to what could be expected in a divinely ordered universe. We need to go further, and we are naturally led to consider the hypothesis that extinctions occur according to a Poisson process.

The Poisson process, or rather the class of Poisson processes, is a most remarkable set of hypotheses, remarkable in their ubiquity, flexibility, and power to represent seemingly disparate phenomena. The basic Poisson process can be visualized in many ways, for example, as a chaotic distribution of events in time. If we accept that 2,230 families have become extinct in the last 562 million years, then if these extinctions occurred according to a Poisson process it would be as if the 2,230 events were randomly distributed, according to a uniform distribution, over that interval. And there is one important mathematical result that would make such a hypothesis plausible: If the families followed independent life histories, then the aggregate process (the Raup-Sepkoski extinction series) should, under mild conditions on those life histories, behave approximately like a Poisson process. This result is a sort of central limit theorem for point processes, and it ensures that the Poisson distribution plays a role in series of events like the role the normal distribution plays in metric data—often as a starting point, sometimes as an ending point.

Thus we have an initial hypothesis, a bit naive perhaps, but as it is based on a first order theory it has at least face credibility. It is also not hard to test. It implies that the number of extinctions X in a stage of duration d has a Poisson distribution with parameter λd, and that the numbers of extinctions in different stages are mutually independent. This would in turn imply that conditional upon the total number of extinctions, 2,230, the distribution of these extinctions among the 76 stages covered by the data set is like the distribution of 2,230 multinomial trials among 76 cells with probabilities proportional to the stage durations. A simple chi-square test of this null hypothesis dispels this notion, however: We find $\chi^2 = 1,565$, on $76 - 1 = 75$ degrees of freedom, inconsistent with such a simple model. But—and this is the central point I wish to make—rejecting a model is the beginning, not the end of an analysis. The question immediately arises, *how* does the model fail? Is the failure necessarily due to the existence of mass extinctions? The argument that led to the consideration of the Poisson model in the first place actually does not quite require the conclusion we arrived at, that is, a simple Poisson process. With but slight changes in the assumptions the argument could lead to a nonhomogeneous Poisson process, for example one whose rate λ varied (perhaps slowly decreased) in time. This enlarges our initial hypothesis, and invites a further test. Are the data consistent with a nonhomogeneous Poisson process with monotonically decreasing rate parameter $\lambda = \lambda(t)$? This question could also be addressed by a chi-square-like test (which would answer "no"), but an inherent limitation of such omnibus tests is that, while they may tell us a hypothesis fails, they do not tell us *how* it fails. So let me try another tack, and take advantage of another remarkable fact about the Poisson distribution. It is that, to a fairly good approximation, if X has a Poisson (λd) distribution, $\sqrt{X}$ has a normal ($\sqrt{\lambda d}$, 1/4) distribution, and thus the square roots of our rates,

$$\sqrt{R} = \sqrt{\frac{X}{d}} \text{ are approximately normal } \left(\sqrt{\lambda}, \frac{1}{4d}\right)$$

If we replace λ by $\lambda(t)$ and note that a slight monotone decrease in $\lambda(t)$ can be captured by modeling $\sqrt{\lambda(t)} = at + b$, we have

$$\sqrt{R} \text{ approximately normal } \left(at + b, \frac{1}{4d}\right)$$

But this is nothing more than a simple linear regression model, with the minor complication that the variances change with d, and the added feature that the variances are known. That is, it suggests that we transform the rates by taking their square roots and regress them on time using weighted least squares (Raup et al. 1983). Looking at the estimated slope should tell us about possible declining extinction; looking at the residuals should tell us about the adequacy of the model. Figure 8.2 shows the transformed data and the fitted regression line.

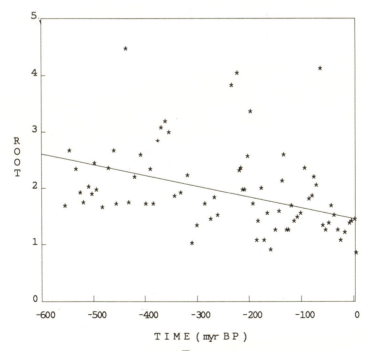

Figure 8.2. Same data as Figure 8.1, but $\sqrt{R}$ versus time, with the weighted least squares regression line.

The results of this analysis are more revealing than the chi-square test. An "eyeball test" applied to this figure suggests that the larger rates are larger than normal scatter would permit, and this visual impression can be verified by more formal methods developed for standard regression models; that is, compared with even this declining rate Poisson process, there is evidence of mass extinctions. For testing for trend, though, we have a dilemma—should we discard the extreme values (the candidates for mass extinctions) or attempt to incorporate them into the analysis? The answer depends upon whether we are more interested in learning about trends in background extinction rates or trends in all extinction rates. If the latter, one possible analysis would be to use a more severe transformation than the square root, namely to regress the logarithm of the rates on time (Figure 8.3) (Quinn 1983).

The effect of using the logarithm of rates is to force all the data into the mold of ordinary normal-theory regression. Note that this is precisely the *wrong* analytical tool to use if we are looking for evidence of mass extinctions, since its primary effect is to destroy any such evidence. Which illustrates one of my constructive principles, namely, that different questions asked of the same data can lead to different models. There can be no such thing as a "neutral" model in such situations, just as data never speak for themselves. Asking about mass extinctions or trends in background Poisson extinctions can lead to models for $\sqrt{R}$; asking about patterns in all extinctions can lead to analyses of patterns in $\log(R)$. And neither of these analyses marks a proper end to the investigation.

I have noted, first, that the argument for a Poisson process of some type is

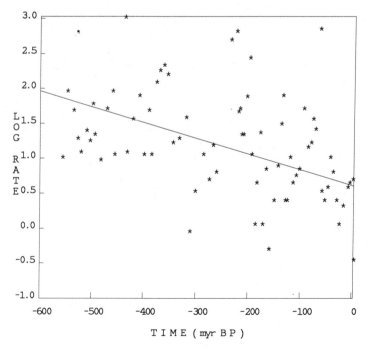

Figure 8.3. Same data as Figure 8.1, but $\log_e(R)$ versus time, with the weighted least squares regression line.

appealing, but that neither a simple constant rate process nor a smoothly declining rate process explains the data. What is happening, evidently, is that the extinction rate has varied much more than is explainable by such simple models. But the fact that such models do not fit the data does not make them uninteresting; rather, they serve as a point of departure for further statistical exploration. If the rate of extinction varies considerably over time, we might first ask what its profile looks like. In particular, can we find strong evidence of two distinct regimes of size of extinction rate, one low (for background extinctions) and one high (for mass extinctions), whether caused by extraterrestrial invasions ("Deathstar May Have Caused Dinosaurs' Demise, Scientists Say"; *Chicago Tribune*, February 21, 1984) or more mundane climatic catastrophes? Or do the data support a continuum of rates, of which the mass extinctions are merely the largest (Figure 8.4)?

For such questions it is fruitful to abandon the approach of inventing a single null hypothesis and then accepting or rejecting it, where acceptance may simply indicate an insensitive test, and rejection a test that is sensitive to an extraneous nuisance factor. It is far better to move to parametric modeling, which permits two models to be fitted within the same framework and their virtues put to a comparative test. Make the data pick between two choices and the data will pick more intelligently than if too much freedom is allowed. The parametric framework I have chosen for the next step is that of modeling each stage's extinction count as a mixture of Poisson distributions. Formally, we might write the distribution of the count of number of extinctions X for a stage of duration d as

$$\text{Prob}(X = k) = \int_0^\infty \frac{e^{-\lambda d}(\lambda d)^k}{k!} \, dG(\lambda)$$

We might think of this as follows: At each stage nature selects a rate λ according to a probability distribution G. Our job is to read the traces and determine G. If we wish, we can think of λ as a latent, unobservable factor representing the sum total of all climatic and astronomical influences, and G the profile or long-term distribution of this factor over time. The single latent factor λ thus captures the correlation of rates within stages. Of course this is an unrealistic model in some respects: Why should nature stick to a rate for a stage and then change abruptly? But this should not be too troubling; we could think of λ as varying more continuously, and the rate for a stage as an average for that stage. If the rate were to spike more abruptly, as with a catastrophe, the stage average would then be higher. Also, we are supposing, perhaps unrealistically, that nature's choice in one stage is independent of the choice in another stage. I will comment more on that later. The main point is that with parametric modeling we may expect (or at least hope) either that such compromises with tractability will be benign or that their effects will be analyzable.

I can report that I have investigated several classes of models for the profile distribution G: two-point mixtures, gamma mixtures (which lead to negative binomial distributions), and mixtures of one-point and gamma distributions (essentially, mixtures of a Poisson and a negative binomial distribution). The methods and models are discussed in greater detail in the Appendix. The choice of these particular models for analysis was motivated by a desire to capture in an analytically tractable framework the broad features of the cases illustrated in Figure 8.4:

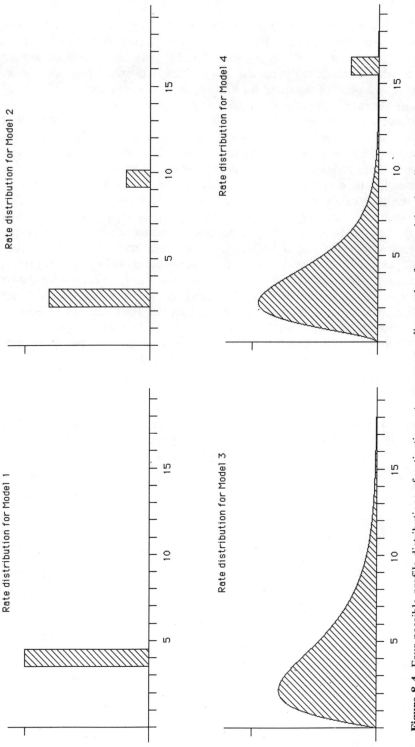

Figure 8.4. Four possible profile distributions of extinction rates, corresponding to the four models described in Table 8.1. From top to bottom these models are model 1, a constant rate; model 2, two discrete regimes; model 3, a gamma distribution of rates; and model 4, a mixture of a gamma distribution and a discrete regime.

a smooth unimodal profile distribution and two types of distributions which incorporate a major "mass extinction" component. Other models could be considered that would achieve this goal, and some might fit the data even better than those considered here. But our primary goal is not to represent the data exactly, it is to force them to choose between these very different types of profile distributions. The models considered are themselves fairly rich and flexible families, and the fact that they are nested allows the application of standard statistical theory in comparing them. The one-point mixture (or Poisson case) is a limiting special case of all of them, and both the two-point mixtures and the gamma mixtures are limiting special cases of the remaining model, the mixed Poisson/negative binomial case.

Since the models have all been fitted by the method of maximum likelihood, any two nested models may be compared by computing the differences in the logarithms of the likelihood functions. The gain that comes with the additional parameters may be compared with a chi-square distribution with degrees of freedom equal to the number of parameters added. The conclusions are summarized in Table 8.1. The main point of interest is the comparison of the models that represent a single regime of extinction rates, models 1 and 3, with those that incorporate a discrete regime of mass extinctions, models 2 and 4. As expected, all of the more complicated models register dramatic improvement over the constant rate Poisson model, model 1. Models 2 and 3 are not nested and hence not directly comparable in a simple way, but both are special cases of model 4. Comparing models 2 and 4 gives a gain of $463 - 363 = 100$ in log likelihood with $4 - 3 = 1$ degree of freedom, compelling evidence that within two-regime models, the two-Poisson mixture is insufficiently rich to capture the variability in the data. The diversity of "normal" extinction rates inherent in the data asserts itself in preferring model 4 to model 2.

The basic question Is there one regime or two? then falls to the comparison of models 3 and 4. Model 3 attempts to capture all of the variability in the rates, including observed mass extinctions, within one unimodal profile distribution with mean $K\delta = 4.2$ extinct families per million years. Model 4, in effect, says that 95% of the time the rates vary in a manner represented by a gamma distribution with mean $K\delta = 3.5$ extinctions per million years, and 5% of the time the rates

Table 8.1. The relative likelihoods for four models for the profile distribution. The numbering of models corresponds to that of the Appendix; models 2, 3, and 4 are like those pictured in Figure 8.4, from top to bottom.

Model	Gain in log likelihood over the Poisson model	Number of parameters
1. Poisson		
($\lambda = 3.96$)	0	1
2. Two-Poisson mixture		
($\lambda = 2.7, \mu = 9.6, \alpha = .2$)	363.5	3
3. Gamma mixture		
($\delta = 2, K = 2.1$)	456.8	2
4. Gamma-Poisson mixture		
($\delta = 1.2, K = 2.9, \lambda = 16, \alpha = .95$)	463.3	4

are in a mass extinction regime of $\lambda = 16$ extinctions per million years. The gain in log likelihood of 6.5 with 2 degrees of freedom corresponds to a P value of about .04. The interpretation of this comparison depends upon how one views the data. On the one hand, a literal interpretation would take this as suggestive (though hardly conclusive) evidence that a mass extinction regime exists. On the other hand, the fragility of such a conclusion should be recognized: It rests basically upon four epochs of mass extinction, and the evidence is sufficiently weak that it could be upset by either minor increases in the estimated durations of these epochs or by minor reassignment of extinctions from these to nearby epochs. Also, we should note that a unimodal "normal" extinction rate profile that dies down only slightly more slowly than the gamma would account for the mass extinction events about as well as does model 4. My own view is that the existence of a separate mass extinction regime should be taken as unproved, but that if indeed such a regime exists, it is present only about 5% of the time. Thus searches for extraterrestrial causal mechanisms (such as phantom "death stars") are not strongly supported by these data, and in any event only causes that operate about 5% of the time should be entertained as consistent with the data. Of course, model 3 already incorporates a wide spectrum of extinction rates, although they vary as a continuum, with larger rates quite possibly arising from the same underlying environmental or biological processes as govern periods of lower extinction rates.

PERIODICITY IN RATES OF EXTINCTION?

In the discussion thus far I have not mentioned any possible periodicity, and it was an apparent 26-million-year periodicity in the Raup-Sepkoski extinction rates that set off the crop of headlines in the past two years, including the May 5, 1985, cover story in *Time* magazine. What is the relationship between the models I have discussed and a possible periodicity in the rates? The simple answer is that there is no necessary relationship. The simple profile distributions considered here could as well arise from periodic as from random causes. The estimates of the parameters of these distributions would be little affected, though the estimated precision of those estimates could be upset more dramatically. I should confess that I am skeptical about the claimed periodicity, however. I accept the tests that have been performed as saying that the rates are not totally random, that they exhibit some serial correlation from stage to stage, although the evidence of even this is not overwhelming. Reassessments of the original Raup-Sepkoski (1984) analysis that attempt to come to grips with selection effects would seem to increase their original P values, which were as low as .0001, to about .03. But, as I have said, rejecting a simple null hypothesis is the beginning, not the end of an analysis. The next step would seem to be to put two alternative parametric models into combat and see if the data express a preference for a periodic hypothesis over a more simple type of serial correlation, such as an autoregressive model that might arise as a consequence of the difficulty of classifying extinctions by stage, and the subsequent problems of anchoring the stages to a time scale. I am now attempting such a comparison with a colleague at the University of Chicago. Some pilot simulations with simple autoregressive models have done well in producing an illusion of periodicity akin to that in the real data. But more is needed, for the stakes are high—if a periodic hypothesis is borne out, we could be relatively safe for another 13 million years; if not, the dangers are more imminent. I do not know what the answer is, but I do know that it will require a combination of good statistical practice and good paleobiology. I am optimistic that the answer will not be long in coming.

APPENDIX

The data used in the example were those analyzed in Raup and Sepkoski (1982) and in Raup et al. (1983). The $n = 76$ rates R_i (where the index i denotes stratigraphic stage) were converted to counts x_i by multiplying by stage duration d_i (in million years). Because the rates had been computed by allocating low-resolution extinctions fractionally among possible stages, many of the x_i values were not integers, and the gamma function $\Gamma(x_i + 1)$ was used in computing $x_i!$ in the models below; no other allowance was made for the fractional allocation. It would be desirable in a fuller analysis to investigate the potential smoothing effect upon the counts of this fractional allocation, although it is doubtful that it would affect the present conclusions greatly.

The models fitted in the present article (in addition to the regression analyses, which are discussed in Raup et al. 1983), are, letting $p(x|d) = \text{Prob}(X_i = x|d_i = d)$

The Poisson Model

$$(8.1) \quad p(x|d) = p(x|d,\lambda) = \frac{e^{-\lambda d}(\lambda d)^x}{x!} \quad x = 0,1,2,\ldots; \lambda > 0$$

The Mixture of Two Poissons

$$(8.2) \quad p(x|d) = p(x|d,\lambda,\mu,\alpha) = (1 - \alpha)\frac{e^{-\lambda d}(\lambda d)^x}{x!} + \alpha\frac{e^{-\mu d}(\mu d)^x}{x!}$$
$$x = 0,1,2,\ldots \quad 0 < \lambda < \mu \quad 0 \leq \alpha < 1$$

The parameter μ may be thought of as the mass extinction rate (if, as here, $\mu > \lambda$), and λ the background rate. If $\alpha = 0$, this model becomes model 1.

The Gamma Mixture of Poissons, or Negative Binomial

$$(8.3) \quad p(x|d) = p(x|d,K,\delta) = \frac{\Gamma(x + K)}{x!\Gamma(K)}(d\delta)^x(1 + d\delta)^{-(x+K)}$$
$$x = 0,1,2,\ldots \quad \delta > 0 \quad K > 0$$

This model corresponds to model 1, but where λ has a gamma distribution

$$g(\lambda|K,\delta) = \frac{e^{-\lambda/\delta}\lambda^{K-1}}{\delta^K\Gamma(K)} \qquad \lambda > 0$$

With this parameterization, the gamma has expectation $K\delta$ and variance $K\delta^2$. This model, and simple ways of estimating the parameters of model 3 by the method of moments, are discussed quite clearly at the beginning of Chapter 4 of Mosteller and Wallace (1984), where the symbol w is used for our d. This gamma distribution may be thought of as a chi-square distribution with $2K$ degrees of freedom, with the variable rescaled by multiplying by $\delta/2$. If $\delta \downarrow 0$ and $K \uparrow \infty$ so that $\delta K \to \lambda$, this model approaches model 1: $p(x|d; K,\delta) \to p(x|d,\lambda)$. So δ may be thought of as a "non-Poissonness" parameter. This model is a prototype of those where λ varies according to a unimodal distribution, such as in Figure 8.4.

The Mixture of a Poisson and a Gamma Mixed Poisson
(Negative Binomial)

(8.4) $\quad p(x|d) = p(x|d,\lambda,K,\delta,\alpha)$

$$= (1 - \alpha)\frac{e^{-\lambda d}(\lambda d)^x}{x!} + \alpha\frac{\Gamma(x + K)}{x!\Gamma(K)}(d\delta)^x(1 + d\delta)^{-x+K}$$

$\quad x = 0,1,2,\ldots \qquad \lambda, \delta > 0 \qquad K > 0 \qquad 0 \leq \alpha \leq 1$

If $\alpha = 0$, this becomes model 1; if $\alpha = 1$, this becomes model 3; if $\delta \downarrow 0$ and $K \uparrow \infty$ and $\delta K \to \mu$, this becomes model 2, so mathematically this model includes the others as special cases. If $\alpha < .5$ and $\delta K > \lambda$, we may think of the negative binomial portion as representing the mass extinctions and λ as the background rate. If $\alpha > .5$ and $\delta K < \lambda$, the negative binomial becomes the background and λ the mass extinction rate.

The models were all fitted by the method of maximum likelihood. For model 1 this is easy: the maximum likelihood estimate of λ is $\hat{\lambda} = \Sigma x_i/\Sigma d_i$, and $\Sigma x_i = 2{,}229.6$, $\Sigma d_i = 562$. For the other models, numerical methods are required. Because the likelihood surfaces in some cases have multiple modes, the simple expedient was taken of evaluating the logarithm of the likelihood function numerically at a grid of values, and exploring its surface interactively. For models 1 to 3 this was quite easy; for model 4 the exploration (in four-dimensional space) was extensive.

The calculation of the log likelihood is a routine matter, using the form

$$\text{log likelihood} = \sum_{i=1}^{n} \log\{\alpha \exp[\log p_1(x_i|d_i)] + (1 - \alpha) \exp[\log p_2(x_i|d_i)]\}$$

in order to avoid overflow, and using Stirling's formula:

$$\log\Gamma(x + 1) \cong \log(2\pi)^{1/2} + (x + .5)\log x - x + \frac{1}{12x} - \frac{1}{360x^3} + \frac{1}{1260x^5}$$

It should be noted that the models fitted are all oversimplified in that I have ignored the apparent decline in average rate, changes in standing diversity, and the effect of errors in determining the time scale. I have fitted several of these models to rates corrected for diversity, but the result (namely that changes in diversity swamp all other variation) suggests that this is not the correct approach, that other changes (in fitness or in the definition of family) have effectively controlled for changes in diversity already.

REFERENCES

Jablonski, D. 1984. Keeping time with mass extinctions. *Paleobiology* 10:139–145.
Mosteller, F. and D.L. Wallace. 1984. *Applied Bayesian and Classical Inference: The Case of the Federalist Papers,* 2nd ed. New York: Springer-Verlag. First Edition, 1964, was entitled *Inference and Disputed Authorship: The Federalist.* Reading, Mass.: Addison-Wesley.
Quinn, J.F. 1983. Mass extinctions in the fossil record. *Science* 219:1239–1240.
Raup, D.M., and J.J. Sepkoski. 1982. Mass extinctions in the marine fossil record. *Science* 215:1501–1503.
Raup, D.M., and J.J. Sepkoski. 1984. Periodicity of extinctions in the geologic past. *Proceedings of the National Academy of Sciences, USA* 81:801–805.
Raup, D.M., J.J. Sepkoski, and S.M. Stigler. 1983. Mass extinctions in the fossil record: Reply. *Science* 219:1240–1241.

Index

Abbott, I., 113, 118
Abele, L. G., 108, 118
Abraham, I., 59, 86
Adaptation, selective, 56 ff
Adaption, limits, 68–69
Adaptive design, 49 ff
Ahistorical universals, 57–58
Alatalo, R. V., 114, 117
Alberch, P., 56, 86
Aleksander, I., 85, 86
Allen, G., 37, 53
Alternative hypotheses, 102
Altruistic traits, 17
Ammonoid genera, 129
Amsterdamski, S., 141, 144
Aoki, K., 17, 18, 21
Arthur, M. A., 123, 131
Arthur, W., 26, 51, 53
Ashburner, M., 64, 84, 86
Assumption, 99 ff
Avifaunas, 116–117
Axel, R., 84, 86
Azumi, Y., 22

Background extinctions, 148–156
Bailey, N. T. J., 43, 53
Bambach, R. K., 137, 144, 146
Bantle, J. A., 59, 86
Bateson, W. E., 32, 37, 53
Beadle, G. W., 38, 54
Behavior
 chaotic, 51
 dynamic, 80
 regulatory, 81–85
 self-organized, 81–85
Berge, C., 64, 86
Biases, 24
Biehl, C. C., 114, 118
Binomial, negative, 157
Biogeography, 109 ff
Biological description, 4
Biological explanation, 3
Biological null models, 100
Biological strategy, 3 ff
Biological theories, 5
Birth-death model, 123
Bishop, J. O., 59, 86, 87

Bonner, J. T., 56, 86
Boolean systems, 80
Bossert, P., 108
Botkin, D. B., 93, 106, 108
Boveri, T., 32, 37, 50, 53
Boveri-Sutton hypothesis, 32, 37
Branching, development, 58
Bretsky, P. W., 126, 127, 128, 131
Bretsky, S. S., 126, 127, 128, 131
Bridges, C. B., 48, 53, 54
Britten, R. J., 59, 84, 86, 87
Brown, D. D., 60, 86
Brown, S., 87
Burns, J. A., 19, 21
Bush, G. L., 62, 86

Cairns, J., 62, 86
Campbell, J., 87
Canalizing ensemble, 81–85
Canalizing system, 80–81
Cardillo, T. S., 87
Carlson, E. O., 37, 48, 53
Carlson, S. S., 89
Carr, T. R., 135, 145
Cartwright, N., 23, 29, 44, 53
Case, S. M., 86
Case, T. J., 113, 117
Castle, W. E., 32, 35, 36, 37, 39, 40, 41, 42, 44, 45, 46, 47, 53
 arguments against, 44 ff
 attack on "linear linkage" model, 37 ff
Catastrophe, complexity, 76–79
Causal model, 28
Chambers, D. A., 60, 80, 89
Chaotic behavior, 51
Charlesworth, B., 57, 86
Cherry, L. M., 89
Chi-square test, 150
Chigusa, S. I., 21
Chikaraishi, D. M., 59, 84, 86
Chromosomes, 36
Cis-acting genes, 59
Clements, F., 93, 106
Codons, RNA, 15
Cody, M. L., 96, 106
Cohen, E., 87
Colonization potential, 116–117

Colwell, R. K., 96, 101, 105, 106, 110, 111, 112, 113, 114, 117, 118
Colwell and Winkler's simulation, 111–114
Community, 93 ff
  ecology, 94 ff
  philosophy, 96 ff
  studies, 93 ff
Competitive interactions, 99
Conjunctive, 49
Connections, regulatory, 64–68
Connectivity, generic, 64–68
Connectivity systems, 80–81
Connell, J. H., 99, 103, 107
Connor, E. F., 114, 115, 118
Constraints, generic, 25
Context independence, 29
Coordination, evolutionary, 79
Corces, V., 59, 86
Cost-benefit condition, 17
Counterfactual false models, 46 ff
Cox, A. V., 144
Crick, R. E., 136, 145
Crow, J. F., 14, 17, 18, 21, 25, 26, 27
Curve-fitting, 25, 29

Dallas, J., 87
Darwin, C., 11, 96, 97, 99, 100
Darwinian gradualism, 19
Darwinian-Mendelian heritage, 56
Data, poor, 114
Davidson, E. H., 59, 84, 86, 87
Deeb, S. S., 86
Deschamps, J., 87
Description, biological, 4
Design, adaptive, 49 ff
Development, inference, 56 ff
Development, branching, 58
Diagram, wiring, 59–62
Diamond, J. M., 96, 106, 110, 114, 115, 116, 118
Dickenson, W. J., 59, 79, 86, 87
Difficulties, methodological, 93
Diffusion model, 14
Disjunctive system, 49
Distribution
  gamma, 157
  null, 100
  species, 111
Diversification
  macroevolutionary, 113–135
  model, 133–139
  Phanerozoic, 133 ff
Diversity
  equilibrium, 126–128
  global, 134
  Phanerozoic, 126–128, 133 ff
  trilobite, 122
Doane, W. W., 59, 86
Domain, predictive, 104
Donehower, L., 87
Dover, G. A., 12, 21, 62, 87
Drift, gene-frequency, 20

DuBois, E., 87
Duhem, P., 141, 142, 144
Dunn, K., 108
Dynamic behavior, 80
Dynamic equilibrium, 121

Ecological community, 94 ff
Ecological generalizations, 94
Ecology, speculative, 96
Edwards, A., 142, 144
Effect, interference, 36
Eldredge, N., 57, 87
Elton, C. S., 94, 107
Engels, W. R., 12, 22
Engineering, piecemeal, 30
Ensemble
  canalizing, 81–85
  evolutionary, 62–64
  genomic, 63
Entrenchment, generative, 27, 51
Equilibrium
  diversity, 126–128
  dynamic, 121
Erdos, P., 64, 87
Ernst, S., 84, 87
Errede, B., 59, 87
Events, major extinction, 123
Evolution
  internal factors, 56
  molecular, 11 ff
Evolution inference, 56 ff
Evolutionary coordination, 79
Evolutionary ensemble, 62–64
Evolutionary morphospace, 128–130
Ewens, W. J., 56, 69, 73, 87
Explanation, biological, 3
Explanatory hierarchy, 5
Explanatory model, 14 ff
Extensive variables, 105
Extinction
  background, 148–156
  events, 121 ff
  major events, 123
  mass, 147 ff
  periodicity, 123–126, 156–157
  Phanerozoic, 138
  rates, 152, 156–157
  taxa, 135
  trilobite, 121–123

False models, 23 ff, 46 ff
  counterfactual, 46 ff
  functions, 30
  use, 28 ff, 46 ff
Falsificationism, 140
Fauna
  equilibrium, 126–128
  fossil, 127
  Ordovician, 126–128
Feigelson, P., 86
Felsenstein, J., 43, 44, 53
Ferson, S., 93, 107

Feyerabend, P. K., 98, 107
Fink, G. R., 62, 88
Fischer, A. G., 123, 131
Fisher, R. A., 13, 14, 21
Fitness, heritability, 26
Flavell, R. B., 12, 21, 62, 87
Flessa, K. W., 138, 144
Fogelman-Soulie, F., 84, 85, 87
Foin, T. C., 132
Fossil
  fauna, 127
  invertebrate, 122
  Lagerstätten, 136, 139, 142
  living, 13
  record, 122
Freese, E., 15, 21
Freud, S., 96
Function
  canalizing systems, 80–81
  incidence, 110–111
  false models, 30
  Haldane, 35, 41
  model, 36

Gadgets, 104 ff
Gamma distribution, 157
Garcia-Bellido, A., 84, 87
Gause, G. F., 95, 107
Gelfand, A. E., 85, 89
Gene expression, 74
Gene frequency drift, 20
Generality, 95
Generalizations ecological, 94
Generative entrenchment, 27, 51
Generic connectivity, 64–68
Generic constraints, 25
Generic genome, 62–64
Generic properties, 26, 62–64, 68–69
Generic states, 72–73
Genes
  cis-acting, 59
  trans-acting, 59
  weakly selected, 19 ff
Genetic code, 13
Genome, generic properties, 62–64
Genomic system
  Boolean, 80
  canalizing, 80–81
  connectivity, 80–81
  ensemble, 63
  regulatory, 59–62
Ghiold, J., 138, 139, 145
Gillespie, D., 62, 87
Gillespie, J. H., 14, 21
Gilpin, M. E., 114, 115, 116, 118
Global diversity, Phanerozoic, 134
Global realism, 23
Glymour, C., 30, 53
Gödel, K., 97, 107
Goldschmidt, R., 37, 47, 48, 50, 53
Goles Chaac, E., 87
Goodwin, B. C., 56, 89

Gotelli, N. J., 110, 114, 116, 117, 118
Gould, S. J., 11, 21, 25, 56, 57, 87, 111, 118, 131, 145
Gradualism, Darwinian, 19
Grant, P. R., 113, 116, 117, 118
Graves, G. R., 110, 114, 116, 117, 118
Green, M. M., 59, 87
Grünbaum, A., 141, 144
Guild, 94

Habitat pools, 116
Hacking, I., 142, 144
Hahn, W. E., 59, 86
Hairston, N., 95, 107, 108
Haldane, J. B. S., 14, 19, 21, 29, 33, 35, 36, 41, 42, 43, 46, 47, 53
Haldane mapping function, 35, 41
Harland, W. B., 137, 144
Harvey, P. H., 109, 111, 115, 118
Hastie, N. D., 59, 87
Hecht, M. K., 4, 7
Helms, C., 59, 88
Hemoglobin, 13
Hempel, C. G., 4, 5, 7, 141, 144
Heritability fitness, 26
Heritage, Darwinian-Mendelian, 56
Heuristic curve-fitting, 29
Hierarchy
  explanatory, 5
  organizational, 4
  taxonomic, 138
Hill, W., 17, 21
Histone, 4, 13
History, verbal paraphrases, 97
Hoffman, A., 4, 5, 6, 7, 8, 135, 136, 137, 138, 139, 140, 142, 143, 144, 145
Holman, E. W., 138, 145
Hood, L., 87
Hough, B. R., 59, 87
Huang, H., 87
Humphreys, T., 58, 87
Humphries, T., 59, 84
Hunkapiller, T., 62, 87
Hutchinson, G. E., 95, 96, 102, 107
Hypothesis, 99 ff
  alternatives, 102
  Boveri-Sutton, 32, 37
  null, 27, 100 ff, 109 ff, 121, 142
  testing, 147 ff

Incidence functions, 110–111
Independence
  context, 24
  quasi, 30
Inference
  development, 56 ff
  evolution, 56 ff
Insight, model producing, 95
Instrumentalist, 39
Integrity, 93
Interactions, competitive, 99
Interference distance, 36

INDEX

Interference effect, 36
Internal factors, evolution, 56
Intuition, 95
Invertebrate fossil record, 122
Island
  comparing, 111
  tables, 110 ff
Iwagami, S., 22
Iwami, M., 22

Jablonski, D., 138, 144, 148, 159
Johnson, C. B., 88
Jukes, T. H., 15, 16, 21

Kacser, H., 19, 21
Karr, A. F., 132
Kauffman, S. A., 20, 25, 26, 27, 53, 59, 80, 83, 84, 85, 87, 88
Kawauchi, Y., 21, 22
Kimura, M., 12, 13, 14, 20, 21
Kin selection, 17
King, J. L., 15, 21
King, R. C., 49, 53
Kitchell, J. A., 135, 138, 145
Kleene, K. C., 59, 84, 88
Kosambi, D. D., 43, 53
Kuhn, T., 96, 97, 98, 107
Kurtz, D. T., 59, 88

Lakatos, I., 142, 145
λ latent factor, 153
Land-bridge, 116–117
Lande, R., 86
Lasker, R. H., 136, 145
Latent factor λ, 153
Levins, R., 29, 31, 52, 53, 95, 105, 107
Levins's prescription, 95
Levy, B. W., 84, 88
Lewis, E. B., 59, 88
Lewontin, R. C., 11, 21, 30, 54, 56, 87, 88
Lidgard, S., 132
Likelihood, maximum, 158
Limits
  adaption, 68–69
  logic, 102 ff
  theories, 104 ff
Lindeman, R. L., 95, 107
Linear linkage model, 37 ff
Linkage, 32, 37 ff
Literature surveys, 103 ff
Living fossils, 13
Llewellyn, P. G., 144
Local realism, 23
Lockley, M. G., 128, 131
Logic, two-value, 102 ff
Low connectivity systems, 80–81
Loya, Y., 94, 107
Lumsden, C. J., 95, 107
Lynch, M., 17, 21

MacArthur, R. H., 93, 95, 96, 107, 109, 126, 127, 128, 131
Macroevolutionary diversification, 133–135

Macroevolutionary model, 133–135
Maguire, B., Jr., 108
Maguire, B., III, 106
Maiorana, V. C., 98, 107
Map distance, 39
Mapping function, Haldane, 35, 41
Margulis, L., 16, 21
Marine animals, 134
Maruyama, T., 14, 21
Marx, K., 96
Mass extinction, 147 ff, 157
Mathematical model, 28
Maximum likelihood, 158
May, R. M., 95, 107, 114, 117, 118
Maynard Smith, J., 95, 107
Mayr, E., 4, 8
McCarthy, B. J., 88
McClintock, B., 59, 88
Mendel, G., 99, 100
Mendelian. See Darwinian-Mendelian heritage
Meselson, M., 86
Methodological difficulties, 93
Mettler, L. E., 21
Mitochondria, 15, 16
Model
  aid to intuition, 95
  best, 24 ff
  biological, 100
  birth-death, 123
  causal, 28
  comparison, 155
  concept, 25 ff
  counterfactual, 46 ff
  diffusion, 14
  diversification, 135–139
  explanatory value, 14 ff
  false, 23 ff, 46 ff
  heuristic curve-fitting, 29
  heuristic value, 14
  linear linkage, 37 ff
  macroevolutionary, 133–135
  mathematical, 28
  null, 100, 149
  phenomenological, 29, 43
  predictive value, 13 ff
  preferred, 47 ff
  producing insight, 95
  regression, 151
  statistical, 111
  stochastic, 6
  template-matching, 36
  templates, 28
  use of false, 28 ff
  usefulness, 17 ff
Modeling, reductionist, 24
Molecular drive, 12
Molecular evolution, 11 ff
Monte Carlo procedure, 130
Monte Carlo simulation, 123
Moore, B., III, 108
Morgan, T. H., 32, 33, 34, 35, 36, 37, 39, 41, 46, 47, 48, 52, 54
Morowitz, H., 106, 108

Morphospace, 121
  evolutionary filling, 128–130
Moscana, A. A., 58, 88
Mosteller, F., 158, 159
Mukai, T., 19, 21
Muller, H. J., 35, 36, 39, 41, 42, 43, 44, 45, 46, 47, 48, 50, 54
  argument, 44 ff
  data, 41
Multiple steady states, 76–74
Mutation, 68–69
  neutral theories, 25
Muto, A., 15, 21, 22
Mycoplasma capricolum, 15

Nagel, E., 97, 107
Nagylaki, T., 14, 21
Negative binomial, 157
Nei, M., 17, 22
Neo-Darwinism, 5–7
Neotropical land-bridge, 116–117
Neutral mutation, 25
Newman, J. R., 97, 107
Nonadditivity, 44
Nonparametric procedure, 148
Non-Poissonness, 158
Null distribution, 100
Null hypothesis, 27, 100 ff, 109 ff, 121, 142
Null models, 100, 149
Null tables, 114

Objectivity test, 99 ff
Odin, G. S., 137, 145
Ohta, T., 12, 22
Operationalist, 39
Oppenheim, P., 4, 5, 7
Ordovician fauna, 126–128
Organization. *See* Self-organization
Organizational hierarchy, 4
Origination
  Phanerozoic, 138
  taxa, 135
Osawa, S., 15, 16, 21, 22

Paigen, K., 59, 88
Painter, T. S., 36, 54
Paleobiology, 121 ff
Panselectionism, 11
Paradigms, 97
Parsimony, 25, 143
Patton, J. L., 86
Peart, D., 132
Pellicer, A., 86
Pena, D., 138, 145
Perfect states, 72–73
Periodicity, extinction, 123–126, 156–157
Peters, R. H., 101, 102, 107
Peterson, P. A., 59, 88
Phanerozoic
  animals, 134
  diversification, 133 ff
  diversity, 133 ff
  extinctions, 138
  originations, 135, 138
Phenomena, rhythmic, 58
Phenomenological model, 24, 43
Philosophy of community studies, 96 ff
Phylogeny, random, 25
Pickton, C. A. G., 144
Piecemeal engineering, 30
Poisson process, 150, 152
Poor data, 114
Popper, K. R., 96, 97, 98, 99, 100, 101, 107, 140, 141, 145
Population
  selection, 69–73
  structure, 18
  subdivision, 17
Pragmatic reductionism, 143
Predation, prudent, 98
Predictive domain, 104
Predictive tests, 47 ff
Predictive value model, 13 ff
Preferred model, 47 ff
Pull of the Recent, 136
Punnet, R. C., 32, 37, 53

Quantitative traits, 19
Quasi-independence, 30
Quine, W. V. O., 141, 142, 145
Quinn, J. F., 138, 145, 152, 159

Raab, P. V., 128, 131
Rachootin, S. P., 56, 88
Rampino, M. R., 123, 131
Random gene-frequency, 20
Random phylogenies, 25
Random walk, 139
Rasmussen, N., 51, 54
Rate
  background extinction, 152
  extinction, 156–157
  mass extinction, 157
Raup, D. M., 25, 26, 27, 111, 118, 122, 123, 124, 125, 129, 131, 132, 136, 137, 138, 143, 145, 146, 148, 149, 150, 151, 156, 157, 159
Realism
  global, 23
  local, 23
  scientific, 23
Reality, 95
Recombination, sexual, 26
Reductionism, pragmatic, 143
Reductionist modeling, 24
Regression model, 151
Regulatory behavior, 81–85
Regulatory connections, 64–68
Regulatory systems
  Boolean, 80
  genomic, 59–62
Reif, W.-E., 4, 8
Rendel, J. M., 56, 88
Renyi, A., 64, 87

INDEX

Research strategy, 3 ff
Revolution, scientific, 97
Rhythmic phenomena, 58
Rich theory, 103, 104 ff
RNA codons, 15
Robertson, A. D. J., 49
Robust relationship, 29
Robust theorem, 31
Roeder, G. S., 62, 88
Root, R. B., 94, 107, 113, 118
Rosenberg, A., 5, 8
Rosenzweig, M. L., 134, 137, 145
Roth, G., 89
Roux, W., 50, 54
Rozanov, A. Y., 137, 146
Russell, B., 97, 108

Salivary gland, 36
Salthe, S. N., 4, 8
Saunders, E., 32, 53
Schaffer, W. M., 51, 54
Schank, J., 26, 51, 54
Schoener, T. W., 95, 99, 103, 107
Schopf, T. J. M., 25, 57, 88, 131, 143, 145, 146
Schultz, G., 86
Scientific realism, 23
Scientific revolution, 97
Scudo, F. M., 95, 107
Seilacher, A., 137, 146
Selection, 56, 68–69
  genes, 19 ff
  kin, 17
  population, 69–73
Selective adaption, 56 ff
Selfish DNA, 12
Self-organization, 56 ff
Self-organized behavior, 80, 81–85
Self-organizing systems, 57–58
Sepkoski, J. J., Jr., 123, 124, 125, 131, 132, 133, 134, 135, 136, 137, 138, 140, 141, 144, 145, 146, 148, 149, 150, 156, 157, 159
Sexual recombination, 26
Shapiro, J. A., 12, 22
Shapiro, R. A., 62, 88
Sheehan, P. M., 135, 141, 146
Sherlock, R. A., 85, 88
Sherman, F., 59, 87, 88
Shymko, R. M., 88
Sidell, R., 113, 117
Significance test, 148, 156
Signor, P. W., III, 132, 136, 140
Silvertown, J. W., 118
Simberloff, D. S., 25, 108, 111, 113, 114, 115, 116, 118, 131, 145, 146
Simon, H. A., 28, 54
Simplicity, 143
Simpson, G. G., 56, 89
Simulation
  Colwell and Winkler's, 111–114
  Monte Carlo, 123

Slatkin, M., 86
Slobodkin, L. B., 94, 95, 96, 98, 104, 105, 106, 107, 108
Smith, A. G., 144
Smith, F. E., 107, 108
Smith, J. M., 82, 89
Smith, M. J., 87
Snelling, N. J., 137, 146
Sober, E., 143, 146
Source pool, habitat, 116
Species
  abundance, 110
  analyses, 110 ff
  distributions, 111
  diversity, 122
  missing, 111
  number, 110
Speculative ecology, 96
Square root transformation, 151
Stadler, L. J., 39, 54
Stanley, S. M., 123, 132
States
  generic, 72–73
  multiple steady, 76–79
  perfect, 72–73
  stationary, 72–73
Statistical models, 111
Statistical tests, 114–115
Steady states, multiple, 76–79
Stearns, S. C., 79, 89
Stigler, S. M., 138, 145, 159
Stirling's formula, 158
Stochastic models, 6
Stothers, R. B., 123, 131
Strachan, T., 87
Stratigraphic boundaries, 137
Strayer, D., 87
Strong, D. R., Jr., 94, 108, 109, 110, 111, 113, 116, 117, 118, 143, 146
Struhl, K., 59, 89
Sturtevant, A. H., 32, 33, 34, 35, 36, 37, 38, 39, 41, 48, 54
Sueoka, N., 15, 22, 86
Sutton. *See* Boveri-Sutton hypothesis
System
  Boolean regulatory, 80
  genomic function, 80–81
  genomic ensemble, 63
  genomic, low connectivity, 80–81
  genomic, regulatory, 59–62
  self-organizing, 57–58
Szyska, L. A., 146

Tautology, 101 ff
Taxa
  extinction, 135
  origination, 135
Taxonomic diversification, 133 ff
Taxonomic hierarchy, 138
Taylor, J. A., 137, 145
Taylor, P. J., 25, 29, 52, 54
Teleology, 11

Teleonomy, 11
Temin, H. M., 12, 22
Template-matching model, 36
Template model, 28
  chi-square, 150
  hypotheses, 147 ff
  objectivity, 99 ff
  predictive, 47 ff
  significance, 148, 156
  weak statistical, 114–115
Testability, 97
Theory, 104
  biological, 5
  limited, 104 ff
  neo-Darwinian, 5–7
  neutral mutation, 25
  paraphrased, 97
  rich, 103, 104 ff
  true, 23 ff
Thistle, A. B., 108, 118
Thomas, R., 60, 61, 89
Thompson, K. S., 56, 88
Tools, 104 ff
Trabert, K., 88
Traits
  altruistic, 17
  quantitative, 19
Trans-acting genes, 59
Transformation, square root, 151
Trick, M., 87
Trilobite
  diversity, 122
  extinction, 121–123
True theory, 23 ff
Two-value logic, 102 ff

Universals, ahistorical, 57–58

Valentine, J. W., 122, 132, 135, 146
Van Valen, L. M., 137, 138, 142, 143, 146

Variables, extensive, 105
Verbal paraphrases of history, 97

Waddington, C. H., 56, 89
Wake, D. B., 56, 57, 89
Wake, M. H., 89
Walker, C. C., 85, 89
Walker, T. D., 135, 146
Wallace, A. R., 97, 99, 100, 108
Wallace, D. L., 158, 159
Walters, R., 144
Weakly selected genes, 19 ff
Webster, C., 56, 89
Weisbuch, G., 87
Weisskopf, V. F., 105, 108
White, T. J., 89
Whitehead, A. N., 97, 108
Whittaker, R. H., 93, 108
Wiane, J.-M., 87
Williams, G. C., 25, 54
Willmer, E. N., 58, 89
Wilson, A. C., 62, 86, 89
Wilson, E. O., 95, 107, 109, 126, 127, 128, 131
Wimsatt, W. C., 11, 12, 23, 24, 25, 26, 27, 29, 30, 31, 51, 52, 54, 55
Winfree, A. T., 58, 89
Winkler, D. W., 101, 106, 110, 111, 112, 113, 114, 117. *See also* Colwell, R. K.
Wiring diagram, 59–62, 69–72
Wittgenstein, L., 97, 108
Woodger, J. H., 96, 97, 108
Wright, S., 19, 20, 22, 51, 55, 98, 108
Wright, S. J., 114, 118

Yamao, F., 15, 21, 22

Ziegler, J. R., 95, 107
Zubay, G., 60, 80, 89